Business Guides on the Go

"Business Guides on the Go" presents cutting-edge insights from practice on particular topics within the fields of business, management, and finance. Written by practitioners and experts in a concise and accessible form the series provides professionals with a general understanding and a first practical approach to latest developments in business strategy, leadership, operations, HR management, innovation and technology management, marketing or digitalization. Students of business administration or management will also benefit from these practical guides for their future occupation/careers.

These Guides suit the needs of today's fast reader.

Gabriel Steinhardt

Data-driven Decision-making for Product Managers

A Primer to Data Literacy in Product Management

Gabriel Steinhardt
Caesarea, Israel

ISSN 2731-4758   ISSN 2731-4766 (electronic)
Business Guides on the Go
ISBN 978-3-031-74663-5   ISBN 978-3-031-74664-2 (eBook)
https://doi.org/10.1007/978-3-031-74664-2

© The Editor(s) (if applicable) and The Author(s), under exclusive license to Springer Nature Switzerland AG 2024

This work is subject to copyright. All rights are solely and exclusively licensed by the Publisher, whether the whole or part of the material is concerned, specifically the rights of translation, reprinting, reuse of illustrations, recitation, broadcasting, reproduction on microfilms or in any other physical way, and transmission or information storage and retrieval, electronic adaptation, computer software, or by similar or dissimilar methodology now known or hereafter developed.

The use of general descriptive names, registered names, trademarks, service marks, etc. in this publication does not imply, even in the absence of a specific statement, that such names are exempt from the relevant protective laws and regulations and therefore free for general use.

The publisher, the authors and the editors are safe to assume that the advice and information in this book are believed to be true and accurate at the date of publication. Neither the publisher nor the authors or the editors give a warranty, expressed or implied, with respect to the material contained herein or for any errors or omissions that may have been made. The publisher remains neutral with regard to jurisdictional claims in published maps and institutional affiliations.

This Springer imprint is published by the registered company Springer Nature Switzerland AG
The registered company address is: Gewerbestrasse 11, 6330 Cham, Switzerland

If disposing of this product, please recycle the paper.

Preface

This book is designed to help product managers understand core concepts related to data, learn to analyze and leverage data, make data-driven product decisions, and communicate data-driven decisions to executive management and engineering.

Data plays a pivotal role in modern product management, enabling product managers with evidence-based decision-making to optimize product feature selection, prioritization, and resource allocation.

Data in product management is compelling evidence, not conjecture and opinion, and it fosters a culture that leverages data to create winning products.

This book is not a guide or manual for data science roles such as Chief Analytics Officer (CAO), Chief Information Officer (CIO), Chief Data Officer (CDO), Chief Digital Officer (CDO), Data Analyst, Data Architect, Data Engineer, and Data Scientist.

This book does not teach artificial intelligence, data lakes, data structures, decision engineering, decision modeling, machine learning, python, SQL, real-time recommendations, Scrum metrics, systems thinking, data analytics, user analytics, and customer analytics.

This book is an introductory primer and reference guide for product managers who wish to use customer feedback, user behavior data, market research, and performance metrics to make informed product decisions.

Caesarea, Israel   Gabriel Steinhardt

Acknowledgments

Without the help and support of some special people, my work and this book would have never become a reality.

Product management practitioners worldwide gave me insights, suggestions, critical reviews, commentary, advice, guidance, and support. A special note of gratitude is extended to them for their invaluable contribution.

My sincere thanks go to all my business partners, fellow instructors, and students worldwide whose feedback and creativity have challenged me. They have candidly shared their thoughts, and I have gained much from each of them. I am grateful for their continued support.

I am also thankful to the ever-professional editorial staff at Springer-Verlag and their partners who have contributed to this literary project. This book was made possible by the diligent work of Prashanth Mahagaonkar, Barbara Bethke, Martina Himberger, Ruth Milewski, and Ramya Prakash.

Finally, I thank my family for their help, commitment, patience, perseverance, faith, support, and love in the life journey they have accompanied me on so far. Without them, I would not be where I am today.

Contents

1	**The Data Age**		1
	1.1 Pre-data Age		1
	1.2 The Data Age		2
	1.3 Agile and Data		2
	1.4 Data Literacy		3
	1.5 Data Science		3
		1.5.1 World Wide Web	3
		1.5.2 Big Data and Data Mining	4
		1.5.3 Data Science	5
		1.5.4 Data Software	5
		1.5.5 Data Science Retention Drill	6
2	**Data-driven Decision-making (DDDM)**		9
	2.1 Qualitative and Quantitative Data		9
		2.1.1 Qualitative Data	9
		2.1.2 Quantitative Data	10
		2.1.3 Qualitative and Quantitative Data Retention Drill	10
	2.2 DDDM and Benefits		11
		2.2.1 DDDM Advantages	11

	2.3 DDDM and Ethics	12
	2.3.1 DDDM Success	12
	2.3.2 Apple Card Example	13
	2.3.3 McKinsey and OxyContin Example	13
	2.4 Data Reliance	14
	2.4.1 Over-reliance on Data	14
	2.5 Intuition-Driven Decision-making	15
	2.5.1 Apple's Product Portfolio	15
	2.5.2 Apple's Product Mishaps	15
	2.5.3 Too Much Data	16
	2.6 Data-driven Versus Intuition-Driven	16
	2.6.1 Balanced Approach	16
	2.6.2 Amazon's Intuition and Data Decision-making	17
	2.7 Privacy and Legal	17
	2.7.1 Privacy Concerns	17
	2.7.2 Apple's ATT	18
	2.7.3 Privacy Regulation	18
	2.7.4 Privacy and Legal Retention Drill	19
3	**Data—Information—Knowledge**	**21**
	3.1 Elements of DDDM	21
	3.1.1 DDDM Components	21
	3.1.2 Bounce Rate Example	22
	3.1.3 Covid-19 Pandemic Example	22
	3.1.4 Elements of DDDM Retention Drill	23
	3.2 Fundamental Statistics	24
	3.2.1 Basic Statistics	24
	3.2.2 Mean (Average)	24
	3.2.3 Mode	25
	3.2.4 Median	25
	3.2.5 Baseline	25
	3.2.6 Correlation	26
	3.2.7 False Correlation Example	27
	3.2.8 Causation	28

		3.2.9	Linear Regression	29
		3.2.10	Fundamental Statistics Retention Drill	30
	3.3	Looking for Patterns		32
		3.3.1	Patternicity and Apophenia	32
		3.3.2	False Positive and Negative Errors	32
		3.3.3	Looking for Patterns Retention Drill	33
4	**The DDDM Process**			35
	4.1	DDDM Introduction		35
	4.2	The DDDM Process		35
		4.2.1	DDDM Steps	35
		4.2.2	The DDDM Process Retention Drill	37
	4.3	Query Formulation		39
		4.3.1	DDDM Query Types	39
		4.3.2	DDDM Query Type Examples	40
		4.3.3	Query Format and Components	40
		4.3.4	Query's Question Component Examples	41
		4.3.5	Descriptive Query (What Has Happened) Example	41
		4.3.6	Diagnostic Query (Why Did it Happen) Example	41
		4.3.7	Predictive Query (What Will Happen) Example	42
		4.3.8	Prescriptive Query (How to Make it Happen) Example	42
		4.3.9	Query Format and Components Retention Drill	42
	4.4	Crafting Questions		43
		4.4.1	Clear Inquiries	43
		4.4.2	Complexity Error Example	44
		4.4.3	Bias or Leading Error Example	44
		4.4.4	Double Negatives Error Example	44
	4.5	Metrics Selection		45
		4.5.1	Choosing Indicators	45
		4.5.2	Metric Categories	45

4.5.3	Adoption Metrics	45
4.5.4	Adoption Rate Example	46
4.5.5	Adoption Rate TFA Example	46
4.5.6	Adoption Metrics Retention Drill	47
4.5.7	Engagement Metrics	48
4.5.8	Engagement Metrics Example	48
4.5.9	Engagement Metrics Retention Drill	49
4.5.10	North Star Metric	50
4.5.11	North Star Metric Unintended Consequences	51
4.5.12	North Star Metric Example (Table 4.1)	52
4.5.13	North Star Metric Retention Drill	53
4.5.14	AARRR Framework	54
4.5.15	AARRR Framework Retention Drill	55
4.5.16	Typical AARRR Metrics	56
4.5.17	Metadata	59
4.5.18	Metadata Examples	59
4.6 Data Inventory		60
4.6.1	Organizing Data	60
4.6.2	Open Data Concept	60
4.7 Data Collection		61
4.7.1	Gathering Data	61
4.7.2	Interviews	61
4.7.3	Observations	62
4.7.4	Focus Groups	62
4.7.5	Questionnaires	62
4.7.6	Online Tracking	63
4.7.7	Online Tracking Specification	63
4.7.8	Event	64
4.7.9	Event Properties	64
4.7.10	User Properties	64
4.7.11	Amazon Online Tracking Example	64
4.7.12	Data Collection Retention Drill	65
4.8 Data Preparation		67
4.8.1	Data Cleaning	67
4.8.2	Data Preparation Retention Drill	68

4.9	Data Analysis	69
	4.9.1 Drawing Insights	69
	4.9.2 Data Analysis Retention Drill	70
4.10	Data Visualization	70
	4.10.1 Graphic Format	70
	4.10.2 Data Visualization Retention Drill	71
4.11	Information Visualization	73
	4.11.1 Trends	73
	4.11.2 Comparisons	74
	4.11.3 Relationships	75
	4.11.4 Revelation Data	75
	4.11.5 Effective Information Visualization	76
	4.11.6 Information Visualization Retention Drill	77
4.12	Communicating Information	79
4.13	Making Decisions	79
	4.13.1 Making Decisions Retention Drill	80

5 Peers and Environment 81
5.1 The Data Analyst 81
    5.1.1 Data Analyst Retention Drill 83
    5.1.2 Data-driven Culture 84
    5.1.3 Attitudes Toward Data 85
    5.1.4 Building a Data-driven Culture 86
    5.1.5 Netflix Data-driven Culture Example 87
    5.1.6 Data-driven Culture Retention Drill 88

Afterword 91

DDDM Glossary 93

About the Author

Gabriel Steinhardt is Blackblot's founder and a recognized international technology product management expert, author, lecturer, and developer of practical tools and methodologies that increase product managers' productivity.

A marketing and information systems MBA with over two decades of experience in product management with technology products, Gabriel has assumed diverse leadership roles with major corporations and start-ups in marketing, product management, and technical undertakings.

Gabriel is the developer of the Blackblot Product Manager's Toolkit® (PMTK) product management methodology, a globally adopted best practice.

1

The Data Age

1.1 Pre-data Age

A lack of continuous data streams and quantifiable metrics marked the era preceding the advent of data-driven decision-making.

During this period, product managers were compelled to rely on their personal experience, intuitive understanding, and a restricted amount of customer data or feedback to guide their decision-making processes.

A primarily self-reliant approach to decision-making, absent of core data, was common among product managers.

Company meetings with internal stakeholders from sales, marketing, engineering, product management, and executive management were a recurrent way to leverage experience instead of data in making decisions.

Yet, making product decisions without data support is risky, particularly in technology and consumer products.

The emergence of masses of data had a transformative impact on decision-making processes in product management.

1.2 The Data Age

The contemporary data age is characterized by the proliferation of data and feedback originating from digital sources.

The surge in data availability has transformed the role of product managers who work with digital products.

With the aid of data analysts who specialize in collecting, organizing, analyzing, and interpreting data, product managers can obtain more accurate product insights from data.

Data analysts are critical in converting raw data into actionable insights. Data analysts inform and enable product managers to make data-driven decisions.

Research has demonstrated that companies with a mature data culture are three times more likely to experience improved decision-making.

Moreover, data-driven decision-making promotes confidence in decision-making, potentially less risk, and increased proactivity.

1.3 Agile and Data

Many modern companies have adopted Agile software development methods, a lightweight iterative development process.

Lightweight iterative development is characterized by continuous plan-develop-test cycles, which enable companies to frequently update and augment their products to address current user needs and build or maintain the product's competitive edge.

In this context, product managers are tasked with and responsible for making data-informed product decisions at each iteration.

Company executives, and product leadership in particular, expect and request that the product manager modify the product feature set and strategy according to the data and insights derived from each iteration.

However, making these product-related decisions can be challenging without robust data.

Successful product managers often validate market need hypotheses as frequently and swiftly as possible. They draw upon various resources, including past experiences, instincts, competitive analyses, and market research.

To ensure accuracy and relevance, even substantive feedback from product users must be validated with objective and accurate data.

Integrating Agile approaches and evidence-based product management constitutes a strategic initiative that can substantially augment a company's competitive advantage and market outcomes.

1.4 Data Literacy

The role of product managers has evolved to necessitate a comprehensive understanding of data concepts, sources, and processes.

This requisite skill is encapsulated in the term "Data Literacy."

Data literacy is the capacity to read, manipulate, analyze, and communicate data.

Data literacy is a conduit for knowledge acquisition, decision-making support, and meaningful communication.

Regarding product management, data literacy has become an indispensable skill and driver of business value.

1.5 Data Science

1.5.1 World Wide Web

The World Wide Web is an information and content-sharing system over the Internet, a global network of computers.

The inception of the World Wide Web, also known as Web 1.0, occurred in the 1990s.

This initial version of the Internet was primarily a platform for making offline information accessible online, mainly via the Web.

Early on, the Internet was characterized by static textual web pages and read-only interaction, offering limited user engagement.

The early 2000s witnessed the emergence of Web 2.0, a significant advancement from its predecessor.

This new version of web technology introduced dynamic web pages, user-generated content, and social media platforms.

A key product feature of Web 2.0 was the introduction of dynamic content, which is online content that adapts according to data, user behavior, and preferences.

During this period, graphic images and early video content became prevalent.

Website usability, interaction, and navigation were greatly enhanced.

Notable milestones during the Web 2.0 era include the launch of Facebook in 2004 and YouTube in 2005. These popular social media platforms, among others, facilitated user interaction, creating vast data sets.

1.5.2 Big Data and Data Mining

The vast data sets created during the Web 2.0 era were termed "Big Data" because their immense size and growing complexity exceeded traditional data-processing software's processing capabilities.

The advent of "Big Data" catalyzed the development of "Data Mining", a data retrieval concept first introduced in 1996.

Data mining is a process dedicated to extracting valuable information from extensive databases. This extraction is facilitated by specific algorithms designed to identify patterns within the data.

However, data mining's progression was initially impeded by the incompleteness of the available algorithms and the limitations of computing power during the 1990s.

The early 2000s saw a significant surge in data mining, attributed to the emergence of improved algorithms and a substantial increase in computing power.

These advancements provided the necessary infrastructure for data mining to thrive and evolve.

1.5.3 Data Science

The growth of big data and improvements in data mining prompted many to utilize data to solve business problems and make informed decisions that enhance product quality and efficiency.

The application of data to solve real-world problems about companies and products was termed "Data Science" and was first introduced by William S. Cleveland in 2001.

Conceptually, data science can be represented by the following formula:

$$\{Data\ Science\} = \{Data\ Mining\} + \{Computer\ Science\}$$

This intuitive formula signifies that data science is a multidisciplinary field combining data mining and computer science methodologies.

The formula underscores the integral role of both data mining and computer science disciplines in data science.

Specifically, within data science, the aspect of making decisions using data was eventually termed "Data-driven Decision-making" and abbreviated as DDDM.

1.5.4 Data Software

Data science has played a central role in developing various categories of software applications, each designed to address specific aspects of data management and utilization.

These categories include:

- **Data Analysis Software (Statistics)**—Data analysis software performs statistical analyses on datasets. Data analysis software provides tools and techniques for conducting descriptive, inferential, and predictive statistical analyses. Data analysis software applications enable users to identify patterns, correlations, and trends within data, facilitating data-driven decision-making processes.

- **Data Visualization Software (Visuals)**—Data visualization software is designed to transform raw data into visual formats such as charts, graphs, and maps. Visual representation of complex data makes it more accessible and understandable, allowing users to quickly grasp insights and trends. Data visualization software is essential for communicating data findings to stakeholders clearly and effectively.
- **Data Dashboard Software (Real-time Metrics)**—Data dashboard software focuses on displaying real-time metrics and Key Performance Indicators (KPIs). Data dashboards provide a centralized interface for monitoring important data points continuously. Data dashboards enable companies to react to changes and make informed operational decisions by presenting real-time data.
- **Comprehensive Data Software (Do It all)**—Comprehensive data software solutions offer an integrated data management approach, encompassing data analysis, visualization, and dashboard software product features. These all-in-one data platforms provide end-to-end capabilities for handling data, from collection and processing to analysis and visualization. Comprehensive data software applications are designed to meet the diverse needs of companies, streamlining workflows and enhancing overall data management efficiency.

These software categories collectively enhance a company's ability to manage, analyze, and visualize data, fostering a data-driven culture and improving decision-making across various business functions.

1.5.5 Data Science Retention Drill

1. What is the title of a skill that entails understanding data concepts, sources, and processes?

    (a) Data Analysis
    (b) Data Literacy
    (c) Data Engineering
    (d) Data Analytics
    (e) Data Knowledge

2. *"Data sets that are very large or overly complex to be handled by traditional data-processing application software."* Which term correctly describes this statement?

   (a) Data Lake
   (b) Data Mining
   (c) Big Data
   (d) Data Science
   (e) Huge Data

3. *"Discovering useful information from large databases data via specific algorithms that look for patterns."* Which term correctly describes this statement?

   (a) Big Data
   (b) Data Lake
   (c) Huge Data
   (d) Data Science
   (e) Data Mining

4. Which is the correct Data Science formula?

   (a) Data Science = Data Mining + Computing Power
   (b) Data Science = Big Data + Computer Science
   (c) Data Science = Big Data + Computer Algorithms
   (d) Data Science = Data Mining + Computer Science
   (e) Data Science = Computer Science + Data Algorithms

Answers: 1-b, 2-c, 3-e, 4-d.

2

Data-driven Decision-making (DDDM)

2.1 Qualitative and Quantitative Data

2.1.1 Qualitative Data

Qualitative data is interpretive and observational and provides descriptive qualities and characteristics that are not quantifiable, defined, or confined by numbers.

Qualitative data includes images, videos, opinions, testimonials, perceptions, and actual behaviors.

Qualitative data is collected through various means:

- **Interviews**—Structured one-on-one question-and-answer sessions where respondents share their experiences and opinions.
- **Focus Groups**—Group sessions with a moderator where participants share their experiences and opinions.
- **Observations**—Recording the behavior and interactions of individuals in a specific environment.
- **Visual Data**—Analyzing images or videos to understand people's experiences or perspectives.

- **Documents**—Analyzing written materials (e.g., newspaper articles, emails, public records) to gain insight into a specific topic.
- **Case Studies**—Analyzing the activities of an individual, group, or company.
- **Online Data**—Analyzing data collected from social media platforms, forums, or online communities to understand views and opinions.

Qualitative data analysis is an interpretive process that provides insights into experiences, perceptions, and behaviors.

2.1.2 Quantitative Data

Quantitative data is characterized by its numerical and factual nature and is expressed using numbers.

Quantitative data is distinguished by its measurable attributes and provides numerical answers to questions such as "How many?" or "How much?"

Quantitative data analysis is a mathematical process that emphasizes numbers, statistics, and other numerical elements, such as the median and standard deviation.

Product managers should analyze both qualitative and quantitative data to make data-driven product decisions.

However, data-driven decision-making (DDDM) primarily focuses on the mathematical analysis of quantitative data.

2.1.3 Qualitative and Quantitative Data Retention Drill

1. Qualitative data is considered to be _____ and _____.
   - (a) Interpretative
   - (b) Mathematical
   - (c) Observational
   - (d) Statistical
   - (e) Calculated

2. Quantitative data is considered to be _____ and _____.

   (a) Statistical
   (b) Observational
   (c) Reasonable
   (d) Numerical
   (e) Factual

3. What type of process is qualitative data analysis?

   (a) Calculated
   (b) Factual
   (c) Interpretative
   (d) Mathematical
   (e) Numerical
   (f) Observational
   (g) Reasonable
   (h) Statistical

4. What type of process is quantitative data analysis?

   (a) Statistical
   (b) Reasonable
   (c) Observational
   (d) Numerical
   (e) Mathematical
   (f) Interpretative
   (g) Factual
   (h) Calculated

   Answers: 1-ac, 2-de, 3-c, 4-e.

2.2 DDDM and Benefits

2.2.1 DDDM Advantages

Data-driven decision-making (DDDM) is a strategic process that prioritizes decisions based on quantitative data instead of intuition or guesswork.

DDDM emphasizes numerical and statistical tools that enhance the understanding of the target market, refine product mix and product features, predict and prepare for future trends, and implement more effective marketing and messaging strategies.

Using data in decision-making processes facilitates more informed and persuasive decisions.

DDDM provides a competitive edge, enables cost reduction, promotes product improvement, and fosters efficiency.

Empirical research indicates that successful technology companies are consistently mining data from their digital products.

Additional research suggests that companies employing DDDM strategies demonstrate superior performance and profitability compared to those that do not.

Data-driven decision-making offers essential and insightful information that guides strategic actions and planning.

DDDM enhances the planning and execution of programs, leading to superior outcomes and improved operational efficiency through streamlined processes.

Moreover, DDDM enables forecasting future trends, allowing companies to prepare proactively for upcoming changes or challenges.

The actionable insights derived from data are concrete and directly applicable to decision-making processes.

2.3 DDDM and Ethics

2.3.1 DDDM Success

Purely relying on data to make business and product decisions can be a proven and successful strategy.

The United States represents one of the most substantial markets for the global airline industry.

Several American airline corporations have adopted a data-driven approach, collecting and analyzing over a hundred customer profile variables while using supercomputers to analyze vast amounts of data to draw insights and act upon them.

For example, the data-driven approach has enabled several major airlines to successfully increase their revenue by 15% annually through offering new services and accurate pricing of their offering.

However, business decisions based purely on data for making business and product decisions can lead to ethical dilemmas, as the data itself lacks ethical considerations.

2.3.2 Apple Card Example

A case in point presenting ethical challenges is the Apple Card, a credit card product developed by Apple and issued by Goldman Sachs. Apple is an American technology company focused on consumer electronics and computer software.

The Apple card, primarily designed for use with Apple Pay (Apple's mobile payment service) on Apple devices, has been controversial due to its algorithm's bias.

The algorithm was found to consistently assign lower credit limits and higher interest rates to women compared to men, prompting an investigation by regulators in New York State.

The Apple Card incident underscores the importance of deploying data ethically.

2.3.3 McKinsey and OxyContin Example

Data-driven decisions should align with a company's mission and values.

Data must be managed appropriately as a crucial business asset, and decision-makers must be adequately trained to utilize it effectively.

Highlighting the ethical implications of data usage is the consulting firm McKinsey and its involvement with Purdue Pharma, a pharmaceutical manufacturer, concerning its OxyContin drug.

In 2013, McKinsey proposed several strategies to drive OxyContin sales, including targeting high-volume opioid prescribers, promoting higher dosages, and attempting to distribute OxyContin directly to patients and pharmacies.

This strategy led to McKinsey settling for nearly US$600 million over its role in the opioid crisis, and Purdue Pharma went into bankruptcy after lengthy litigations and claims totaling US$10-US$12 billion.

In conclusion, while data-driven decision-making (DDDM) can offer significant benefits, it is not free from moral considerations.

Therefore, DDDM must be implemented with an ethical mindset to prevent potential harm.

2.4 Data Reliance

2.4.1 Over-reliance on Data

Over-reliance on data can often prove to be counterproductive.

This counter-productivity is exemplified in stock market predictions, where algorithms are employed to forecast security prices.

Despite the precision of the mathematical models, which base their predictions on historical performance, market outlook, and current security information, the mathematical models have repeatedly failed to anticipate stock market crashes.

The inability to foresee stock market crashes underscores the limitations of data-driven predictions, as recorded history is replete with instances where financial markets experts, armed with accurate market data and sophisticated mathematical models, were unable to envision market downturns.

Moreover, the imposition of pre-defined numerical targets can inadvertently suppress innovation, risk-taking, and creativity; attributes that are highly prized in most professional environments.

Over-reliance on data may prioritize short-term objectives at the expense of long-term goals, thereby undermining broader strategic aspirations.

Thus, balancing data reliance and pure human intuition is crucial to achieving optimal outcomes.

2.5 Intuition-Driven Decision-making

2.5.1 Apple's Product Portfolio

Steve Jobs, the renowned founder of Apple Computer, was a prominent figure who questioned the reliance on market research and historical data.

Steve Jobs was known for his belief that consumers are often unaware of their needs, advocating instead for a reliance on intuition over conventional market research and data analysis.

During his tenure as Apple's CEO, Steve Jobs' unique intuitive approach to decision-making led to a paradigm shift in Apple's product portfolio.

This shift resulted in the launch of a series of groundbreaking products, such as the iPad, iPhone, and MacBook, which have since become synonymous with innovation in the technology sector.

However, it is essential to note that an intuition-driven approach to decision-making is not without its pitfalls.

2.5.2 Apple's Product Mishaps

Apple's history includes several notable product failures, which are cautionary tales.

For instance, the Apple Newton Personal Digital Assistant (PDA), launched in 1993, was a commercial failure due to its faulty graffiti functionality (handwriting recognition), a critical product feature for a PDA.

Similarly, the Apple Lisa computer, released in 1983, failed to gain market traction due to its high price point, the versatility of IBM-PC clones, and the cannibalization of the Apple Lisa sales by the more advanced and compact Apple Macintosh.

Other examples include the Apple MobileMe cloud computing service, plagued by bugs and reliability issues, and the Apple G4 Cube, discontinued within a year of its launch due to its high cost and lack of an included monitor.

2.5.3 Too Much Data

Data can be overwhelming.

The appeal of intuition-driven decision-making often lies in its simplicity, particularly when faced with the daunting task of data analysis.

Understanding and interpreting data can be complex and time-consuming, leading individuals to seek more manageable alternatives.

Intuition-driven decision-making offers a quicker and more cost-effective solution.

However, it is crucial to recognize that intuition alone is insufficient for robust decision-making.

2.6 Data-driven Versus Intuition-Driven

2.6.1 Balanced Approach

Empirical evidence suggests that data and algorithmic analysis often outperform human intuition across various scenarios.

Therefore, a balanced approach that combines intuition with data-driven insights is recommended for optimal decisions.

Empirical studies have revealed that intuition, instead of data-driven insights, predominantly influences a significant proportion of managerial decisions.

Other studies suggest that at least half of the routine business decisions within these companies are based on intuition without any empirical evidence.

Decisions primarily guided by emotions and intuition tend to be reactive rather than proactive.

In the contemporary business landscape characterized by abundant invaluable data, relying solely on intuition for decision-making could be imprudent.

The importance of substantiating decisions with data cannot be overstated, particularly in an era where data is heavily used to make business decisions.

Yet, decision-makers lose their business instincts when they focus too much on data.

Therefore, it is imperative to balance intuition and data in decision-making, ensuring that decisions are both informed and intuitive.

2.6.2 Amazon's Intuition and Data Decision-making

Jeff Bezos, the founder of Amazon, employed a balanced approach of intuition and data-driven decision-making. Amazon is an American eCommerce and technology company.

During his tenure at D.E. Shaw, a hedge fund, Bezos was tasked with exploring new business opportunities on the Internet, which held immense potential in the early 1990s.

During his research, Bezos encountered data indicating an unprecedented growth rate in the usage of the World Wide Web.

After careful consideration, Bezos identified books as the most feasible product that could be sold online.

Creating an online bookstore with millions of book titles appealed to Bezos.

Bezos decided to bet on the Internet, and in 1994, he founded Amazon as an online book store.

Although data about the World Wide Web was available, online bookstores were non-existent then.

Bezos combined the available web data with his intuition about the future of global shopping.

This blend of data and intuition culminated in creating Amazon, an eCommerce company that has since become a household name worldwide.

2.7 Privacy and Legal

2.7.1 Privacy Concerns

People are highly protective of their private and personal data, especially regarding their location, financial transactions, biometric and medical data, and children.

Most consumers, over 90%, believe that companies should be transparent about what they do and how they use user data.

There are growing concerns about how companies and governments might misuse personal data, such as selling it or using it inappropriately.

These concerns have led to a privacy industry that offers services like anonymous web browsing proxies, Virtual Private Networks (VPNs) for encrypted traffic and location concealment, private browsers for ad-free and untracked web surfing, and encrypted email services for private communication.

Some businesses collect, store, and sell users' private information found online, while other companies offer services to remove such information from the Internet.

2.7.2 Apple's ATT

As awareness about privacy, ethical, and social issues related to data increases, companies are being questioned about their data collection, storage, and usage practices.

For instance, many iPhone apps have been found to quietly gather and share user data, such as phone numbers and locations.

The iPhone, developed by Apple, is a smartphone that integrates the functionalities of a handheld computer, media player, digital camera, and cellular phone into a single device with an intuitive touchscreen interface.

In response to users' privacy concerns, Apple introduced the App Tracking Transparency (ATT) product feature in iOS (iPhone's operating system), which allows users to control whether a mobile application can track their activity across other mobile applications and websites for advertising or data brokerage purposes.

2.7.3 Privacy Regulation

To address privacy concerns, governments and regulators worldwide have enacted privacy laws.

The European Union enacted the General Data Protection Regulation (GDPR) to manage user information privacy, while in the U.S., antitrust laws are used to protect user privacy.

In the U.S., antitrust laws also prevent large companies or conglomerates from gaining unfair market power by collecting and sharing user data internally among themselves.

For example, Alphabet, the parent company of Google (Search engine), Gmail (web-based email service), and YouTube (video-sharing platform), can access and share user data within its ecosystem of owned companies and external partners.

There have been several instances where U.S. regulators have taken action on user data privacy concerns.

In 2019, Facebook (social media company) was fined US$5 billion by the Federal Trade Commission (FTC), the U.S. federal civil antitrust law enforcement agency, for sharing users' political information with Cambridge Analytica (British political consulting firm).

In 2020, Zoom (video telephony software) was found to be data-mining user conversations and sharing the data with LinkedIn (business social media platform).

In 2021, Facebook fired 52 employees for misusing their access to user data, including male employees who obtained location data on women they were interested in.

The U.S. Senate's Judiciary and Commerce committees periodically discuss personal privacy issues, highlighting the balance between the right to privacy and the willingness to disclose it.

Product managers need to understand the legal aspects of data in their jurisdiction and, as a general rule, should avoid their products collecting personally identifiable data unless required by law.

2.7.4 Privacy and Legal Retention Drill

1. What did Apple's App Tracking Transparency (ATT) product feature do?

    (a) Improved app performance
    (b) Increased app download rates
    (c) Enhanced user privacy control
    (d) Streamlined app development processes

2. What does GDPR stand for?

    (a) Group Data Privacy Region
    (b) Global Data Privacy Region
    (c) Graduate Data Program Report
    (d) General Data Privacy Regulation
    (e) General Data Protection Regulation

3. To which entity does the GDPR belong?

    (a) Soviet Union
    (b) European Union
    (c) United States of America
    (d) North Atlantic Treaty Organization (NATO)

4. Which entity uses antitrust laws to protect user privacy?

    (a) China
    (b) Soviet Union
    (c) European Union
    (d) United States of America

Answers: 1-c, 2-e, 3-b, 4-d.

3

Data—Information—Knowledge

3.1 Elements of DDDM

3.1.1 DDDM Components

The data-driven decision-making (DDDM) process comprises four integral components: data, information, knowledge, and decision/s.

- **Data** is a collection of unorganized facts. Data, whether numerical or textual, lacks any specific meaning or context.
- **Information** is organized data. Information has meaning and context that can be useful and can be understood.
- **Knowledge** is the capacity to comprehend and utilize information, forming the foundation for making decisions. Knowledge encompasses the insights and conclusions derived from the information.
- **Decision/s** are judgments or choices that instigate action based on acquired knowledge. Decisions are determinations reached after consideration.

Consequently, the DDDM process involves collecting raw **data**, organizing the data into meaningful **information**, acquiring **knowledge** by identifying insights and drawing conclusions, and making **decisions** that lead to the actions.

3.1.2 Bounce Rate Example

The DDDM process and its components can be demonstrated with an example of the "Bounce Rate", a metric utilized in web traffic data analysis.

The bounce rate denotes the proportion of website visitors who access a specific web page and exit ("Bounce") without exploring other pages on the same website.

A high bounce rate is unfavorable, suggesting a need for improvements to draw and engage visitors and motivate them to navigate the website further.

In the context of DDDM, the bounce rate instances collected on the website server constitute the **data**.

For example, a statistical analysis yielding a bounce rate of 70% represents the **information**.

A conclusion that the high bounce rate is attributable either to the wrong audience visiting the website or to some flaw on the website causing the target audience to leave constitutes **knowledge**.

The determination to modify the website's design to enhance its appeal and optimize website traffic from search engines represents the **decision**.

3.1.3 Covid-19 Pandemic Example

Another example of the data-driven decision-making (DDDM) process is in the context of the Covid-19 pandemic.

In this scenario, medical records of Coronavirus patients that are gathered and preserved are the **data**.

Statistical analysis of the Coronavirus patients' medical records reveals that individuals over the age of 60 tend to suffer from severe and acute symptoms of the disease, is the **information**.

Recognizing that Coronavirus patients who are over 60 years old are at a greater risk of experiencing severe health complications and symptoms is the **knowledge**.

The vaccination strategy that prioritizes vaccinating adults over 60 years old against the Coronavirus is the **decision**.

3.1.4 Elements of DDDM Retention Drill

1. Which statement about **Data** is incorrect?

   (a) Data does not carry any specific meaning.
   (b) Data can be numbers or words.
   (c) Data is unorganized facts.
   (d) Data has meaning and context.

2. Which statement about **Information** is incorrect?

   (a) Information is useful and can be understood.
   (b) Information is the resolution of uncertainty.
   (c) Information is organized data.
   (d) Information has meaning and context.

3. Which statement about **Knowledge** is incorrect?

   (a) Knowledge is the foundation for making decisions.
   (b) Knowledge encompasses insights and conclusions.
   (c) Knowledge enables manipulation and sorting of raw data.
   (d) Knowledge is the capacity to comprehend and utilize information.

4. Which statement about **Decisions** is incorrect?

   (a) Decisions are determinations following consideration.
   (b) Decisions are judgments or choices that instigate action.
   (c) Decisions are insights and conclusions based on information.
   (d) Decisions are based on acquired knowledge.

   Answers: 1-d, 2-b, 3-c, 4-c.

3.2 Fundamental Statistics

3.2.1 Basic Statistics

Fundamental statistics refers to the basic principles and techniques of organizing and analyzing data.

Statistics help make sense of the world by producing numerical information and interpreting the results.

Product managers often utilize key statistical measures such as mean, mode, median, correlation, causality, and linear regression.

Mean, mode, and median are the most commonly used statistic tools, and they all represent a number that expresses the central or typical value in a data set.

3.2.2 Mean (Average)

The mean, often called the average, represents the central value of a numerical dataset.

The mean is computed by summing all the numbers in the dataset and then dividing the sum by the total count of values.

For example, a dataset contains three values: 6, 18, and 24. The sum of these values is 48. When this sum (48) is divided by the total number of values in the dataset (3), the result is 16. Hence, the mean (or average) of this dataset is 16.

However, the mean may not always provide the most accurate representation of data, especially in large datasets with many outliers (values that significantly deviate from the others) or skewed data distribution.

In such cases, the mean might not lead to the best decisions in statistical analysis.

Therefore, the mean isn't always the most reliable measure for decision-making when dealing with numerous outliers.

3.2.3 Mode

In statistics, the most frequent number in a given data set is mode.

For example, a dataset includes these values: 16, 17, **36**, **36**, 57, 65, 67, 74, 78, 79, and 91.

Mode is the most common number in a dataset, which this particular dataset is 36, as it occurs twice, a frequency higher than any other number.

3.2.4 Median

The median is the middle value in a sorted data set. The middle value splits a sorted data set into two equal parts.

For example, a dataset includes these values: 16, 17, 36, 36, 57, **65**, 67, 74, 78, 79, and 91.

The data points are sorted in ascending order to find the median, and then the middle number is identified.

The median of this dataset is 65 because it is the middle value of the sorted dataset.

Like mean and mode, the median provides a number expressing the central tendency or a typical value in a data set.

The median is a measure of the center of a data set and can help describe how data values are distributed and how they compare to each other.

The median is beneficial when the dataset is skewed or has outliers, which can affect the mean.

3.2.5 Baseline

A baseline refers to the initial value of a variable that serves as a basis for comparison.

A baseline is established by measuring a variable of interest at a preliminary stage and then comparing it with subsequent measurements to examine the changes.

In product management, a baseline is used to evaluate the performance of product features or a product, guiding the decision-making process.

The baseline value is the starting point to track product performance over time and measure product decisions' effectiveness.

A baseline can also compare a product with its competitors or industry standards.

In software applications, a baseline value often serves as a reference point for performance measurement. For instance, the processing time required to generate a report or fulfill a query.

First, an initial value (the baseline) is established. Subsequently, any changes or optimizations are implemented, and the processing time is measured again.

The impact of the changes is assessed by comparing the updated processing time to the baseline value.

3.2.6 Correlation

In statistics, correlation quantifies (measures) how two variables are related to each other.

Correlation describes a relationship or connection between two or more variables.

A correlation coefficient is a number between −1 and 1 that indicates the strength and direction of the relationship.

There are three types of correlation: positive, negative, and zero (Fig. 3.1).

A positive correlation means that both variables change in the same direction. For example, when user satisfaction levels increase, product revenue increases.

A negative correlation means that the variables change in opposite directions. For example, user satisfaction levels may decrease if a software product's bugs increase.

A zero correlation means no linear relationship exists between the variables. For example, there is no linear relationship between the number of colors used in a product and the product's durability.

CORRELATION TYPES

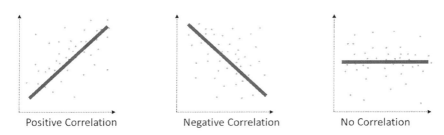

Fig. 3.1 Correlation types

However, correlation and causation are separate concepts, and the fact that two variables are moving in the same direction does not mean that they are related or that one variable is causing movement in the other variable.

It is essential to understand that correlation and causation are different, and product managers analyze data to seek and discover genuine causal relationships and avoid false correlations, which are mistaken for causation.

3.2.7 False Correlation Example

A false correlation is a statistical phenomenon where two variables appear to be related, but in reality, they are not.

A false correlation can occur due to chance, confounding variables (third influential variable), or random associations.

An example of a false correlation is the relationship between the average personal expenditure on ice cream and cellular minutes in the summer (Fig. 3.2).

People spend more money on ice cream and use more cellular minutes during the summer.

However, this does not mean that consuming ice cream causes people to talk more on the phone or vice versa.

The reality is that people talk more on their cellular phones during the summertime because they are on vacation, have more free time to

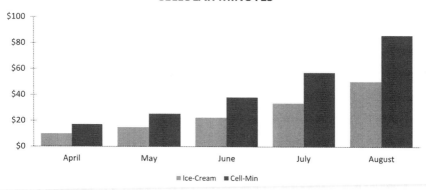

Fig. 3.2 False correlation

connect with friends and family, and consume more ice cream because the temperatures outside are higher.

Accordingly, a correlation may be purely coincidental and not reflect any meaningful relationship.

3.2.8 Causation

Causation describes the relationship between cause and effect between two variables.

Causation in product management means a direct and consistent effect of one variable on another, such as a product feature or a user behavior.

For example, adding an automated Chabot to a website will drive away website visitors seeking support and interaction with a live person.

To establish causation for product management, experimentation and tests are required to isolate the impact of one variable on another while controlling for other factors that may influence the outcome.

Ultimately, causation attempts to uncover that one data point will consistently cause another data point to happen (Fig. 3.3).

For example, progressively higher outside temperatures cause people to consume more ice cream frequently.

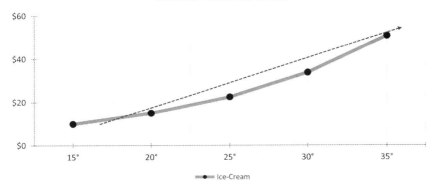

Fig. 3.3 Causation

3.2.9 Linear Regression

Linear regression is a statistical technique to estimate the magnitude and direction of the association between one outcome variable (typically denoted by Y in formulas and graphs, and also referred to as the "dependent variable") and a set of predictor variables (typically denoted by X in formulas and graphs, and also referred to as the "independent variables").

In statistical terms, linear regression refers to the relationship between an outcome variable (the data of interest) and a predictor variable (the data used to forecast the outcome variable).

Linear regression also shows the relationship between dependent and independent variables:

- The dependent variable is the data that will be measured.
- The independent variable is the data used to predict the dependent variable.

Linear regression is used in predictive analysis to illustrate how one variable influences another variable or how variations in one variable induce alterations in another, implying causality (Fig. 3.4).

Product managers explore for causation and employ linear regression for prediction and manipulation.

LINEAR REGRESSION

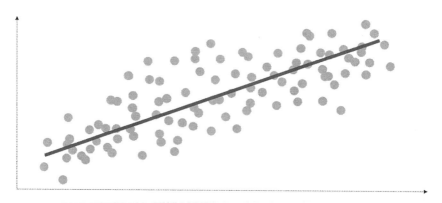

Fig. 3.4 Linear regression

For example, product managers will plot a linear regression to analyze and predict if adding a specific type of product feature will increase user satisfaction.

3.2.10 Fundamental Statistics Retention Drill

1. What is the definition of the mean (average)?

    (a) The middle value in a sorted dataset.
    (b) The most frequently occurring value in a dataset.
    (c) The sum of all values divided by the number of values.
    (d) The sum of maximum and minimum values.

2. What is the mode in a dataset?

    (a) The average value.
    (b) The most frequent value.
    (c) The middle value.
    (d) The most conspicuous value.

3. What is the median in a dataset?

    (a) The middle value.
    (b) The average value.

(c) The most frequent value.

(d) The largest value.

4. What does the term "Baseline" refer to in statistical analysis?

    (a) The average value of a dataset.
    (b) The most extreme value in a dataset.
    (c) The initial value or reference point for comparison.
    (d) The sum of all values divided by the number of values.

5. What does correlation measure?

    (a) The strength of a linear relationship between two variables.
    (b) The average value of a dataset.
    (c) The difference between the maximum and minimum values.
    (d) The causation factor between two variables.

6. What is a false correlation?

    (a) A strong positive relationship between two variables.
    (b) A weak negative relationship between two variables.
    (c) A manufactured relationship between two variables.
    (d) A coincidental association between two variables.

7. What is causation?

    (a) A statistical measure of association between two variables.
    (b) A relationship that appears to be occasional but is not.
    (c) A set of observational data without experimental evidence.
    (d) A cause-and-effect relationship between two variables.

8. What is Linear Regression?

    (a) A statistical method used for classification tasks.
    (b) A model that predicts a continuous output based on input product features.
    (c) A technique to find the median of a dataset.
    (d) A measure of association between two categorical variables.

Answers: 1-c, 2-b, 3-a, 4-c, 5-a, 6-d, 7-d, 8-b.

3.3 Looking for Patterns

3.3.1 Patternicity and Apophenia

The volume of data and information that people encounter daily is immense.

To cope with data overload, the human cognitive system is predisposed to detect patterns in the data.

This phenomenon is known as Patternicity or Apophenia, which can pose a significant challenge for data analysis.

Patternicity or Apophenia is the human propensity to seek patterns in random data.

Seeking patterns devoid of logic and validity happens in business or gambling.

For example, a day trader who engages in day trading, a highly risky and speculative type of securities selling and buying, who has had several good trades while dressed in their pajamas, may now begin to trade regularly while in their pajamas.

While the human cognitive system tends to detect patterns in data, it cannot comprehend probabilistic reasoning.

Moreover, human brains are ill-equipped to factor randomness and small data sets into decision-making.

3.3.2 False Positive and Negative Errors

In the attempt to look for patterns, false positive and false negative errors occur.

False positive and false negative errors are two types of mistakes that can occur in statistical decision-making.

A false positive error is when a non-existent pattern is incorrectly detected. This error means assuming a relationship between data variables when there is none.

A false negative error is when an existing pattern is undetected. This error means a present relationship between data variables goes undiscovered.

A false positive error could imply the development of unnecessary product features, burdening the company with cost and effort for product features that offer no value to the user.

A false negative error could entail not building essential product features, thus exposing the company to a competitive disadvantage and missed business opportunities.

Product managers must look for patterns everywhere yet validate the patterns with ample data with reasonable statistical confidence.

Product managers are responsible for identifying patterns across various domains and verifying them with ample data and reasonable statistical confidence.

3.3.3 Looking for Patterns Retention Drill

1. What is Patternicity or Apophenia?

    (a) The ability to comprehend probabilistic reasoning.
    (b) The process of detecting existing deterministic patterns.
    (c) The tendency to seek patterns in random information.
    (d) The ability to identify patterns across various domains.

2. Which of the following best describes a false positive error?

    (a) Missing an existing pattern.
    (b) Detecting a non-existent pattern.
    (c) Comprehending probabilistic reasoning.
    (d) Data variables go undiscovered.

3. What does a false negative error imply?

    (a) A non-existent pattern is incorrectly detected.
    (b) Probabilistic reasoning is well-understood.
    (c) Adding product features that offer no value.
    (d) An existing pattern is undetected.

4. What responsibility do product managers have regarding patterns and statistical confidence?

   (a) Validate patterns with ample data and reasonable statistical confidence.
   (b) Alerting executives to any competitive disadvantage and missed business opportunities.
   (c) Intuitively develop most product features.
   (d) Look for patterns everywhere without validation.

Answers: 1-c, 2-b, 3-d, 4-a.

4

The DDDM Process

4.1 DDDM Introduction

Data-driven decision-making (DDDM) is a systematic process that leverages data and statistical analysis to identify patterns and gain insights that support business and product decisions.

The DDDM process involves collecting targeted, high-quality data, analyzing it using statistical methods and data visualization techniques, and deriving actionable insights to make informed business, strategic, and product decisions.

The premise of the DDDM process is in its structure and objectivity.

4.2 The DDDM Process

4.2.1 DDDM Steps

The data-driven decision-making process comprises several critical steps to ensure accurate and relevant data-informed decisions.

1. **Query Formulation**—Creating precise requests for data retrieval and analysis.
2. **Crafting Questions**—Designing clear, relevant inquiries to guide data exploration.
3. **Metrics Selection**—Choosing specific performance indicators for evaluation.
4. **Data Inventory**—Cataloging available data sources.
5. **Data Collection**—Gathering raw data from different sources.
6. **Data Preparation**—Cleaning, transforming, and structuring data for analysis.
7. **Data Analysis**—Extracting patterns and insights from data, converting data to information.
8. **Data Visualization**—Visual/graphical representations of numerical statistical figures.
9. **Information Visualization**—Visual/graphical representations of insights and information in support and context of decision-making.
10. **Communicating Information**—Presenting findings effectively to stakeholders.
11. **Making Decisions**—Using data-driven insights to make informed choices.

The DDDM process begins with **Query Formulation**, where precise data retrieval or analysis requests are created.

Query formulation is followed by **Crafting Questions**, which involves designing clear and relevant inquiries to guide data exploration.

Next is **Metrics Selection**, where specific performance indicators or measures for evaluation are chosen, setting the stage for targeted analysis.

Data Inventory involves cataloging available data sources and understanding their characteristics, which is crucial for effective data management.

The process continues with **Data Collection**, where raw data is gathered from various sources, and **Data Preparation**, which involves cleaning, transforming, and structuring the data for analysis.

During the **Data Analysis** phase, patterns and insights and from data, converting data to information.

With specialized software, **Data Visualization** is created to graphically and visually represent numerical statistical figures.

The following step is **Information Visualization**, where charts and graphs are created to enhance the understanding of insights and information derived from the data, supporting the decisions that must be made.

The next step is **Communicating Information**, presenting findings effectively to stakeholders, and getting their input.

The process concludes in **Making Decisions**, where data-driven insights guide making informed choices.

The comprehensive DDDM process ensures that decision-making is informed and strategic, leveraging the full potential of available data.

4.2.2 The DDDM Process Retention Drill

1. In DDDM, which best describes Query Formulation?

    (a) Cleaning, transforming, and structuring data for analysis.
    (b) Constructing broad questions that tap available data sources.
    (c) Choosing specific performance indicators for evaluation.
    (d) Creating precise requests for data retrieval and analysis.

2. In DDDM, what is the primary objective of Crafting Questions?

    (a) Building a research foundation that drives product development.
    (b) Extracting patterns and insights from gathered data.
    (c) Designing clear, relevant inquiries to guide data exploration.
    (d) Gathering stakeholders' input and cataloging their interests.

3. In DDDM, what is the purpose of Metrics Selection?

    (a) Choosing specific performance indicators for evaluation.
    (b) Using data-driven data to make informed product decisions.
    (c) Designing statistical parameters around available data sources.
    (d) Convincing data analysts to adopt to evidence-based methodologies.

4. In DDDM, which statement best defines Data Inventory?

   (a) Cataloging available data sources.
   (b) Gathering raw data from different data sources.
   (c) Systematically organizing user data sets.
   (d) Presenting a centralized repository of information.

5. In DDDM, what is Data Collection?

   (a) Cleaning, transforming, and structuring data for analysis.
   (b) Techniques and tools used to acquire data insights.
   (c) Gathering raw data from different sources.
   (d) Measuring the impact of targeted statistical manipulation.

6. In DDDM, what is involved in Data Preparation?

   (a) Standardizing post-analysis data sets and data errors.
   (b) Cleaning, transforming, and structuring data for analysis.
   (c) Enriching, enhancing, and expanding missing data with insights.
   (d) Converting raw data into information.

7. In DDDM, what does Data Analysis entail?

   (a) Extracting patterns and insights from data, converting data to information.
   (b) Organizing complex data sets with statistical techniques.
   (c) Testing data hypotheses and extracting data outliers.
   (d) Joining information and data while simultaneously drawing conclusions.

8. In DDDM, what is the focus of Data Visualization?

   (a) Presenting findings effectively to stakeholders.
   (b) Visual/graphical representations of numerical statistical figures.
   (c) Putting charts, graphs, maps, and other visual tools to use.
   (d) Extracting meaningful and actionable insights from graphics.

9. In DDDM, what best describes Information Visualization?

   (a) Storing large volumes of visual data in a database for long-term archival purposes.
   (b) Creating random visual designs without any data context or analysis for rapid accessibility.

(c) Visual/graphical representations of data using numerical signals represent data insights.
(d) Visual/graphical representations of insights and information in support and context of decision-making.

10. In DDDM, what is the objective of Communicating Information?

    (a) Ensuring that stakeholders' intuition is included in product decisions.
    (b) Presenting findings effectively to stakeholders.
    (c) Demonstrating data complexities and clarifying them to stakeholders.
    (d) Avoiding overwhelming decision-makers with information.

11. In DDDM, what does Making Decisions involve?

    (a) Leveraging anecdotal evidence to select product features.
    (b) Using information to support business case studies.
    (c) Using data-driven insights to make informed choices.
    (d) Organizing accurate and relevant data to make product decisions.

Answers: 1-d, 2-c, 3-a, 4-a, 5-c, 6-b, 7-a, 8-b, 9-d, 10-b, 11-c.

4.3 Query Formulation

4.3.1 DDDM Query Types

Query formulation is creating precise requests for data retrieval or analysis.

In DDDM, queries are classified according to a timeline that addresses the past, present, and future.

There are four types of DDDM queries:

- **Descriptive Query (Past)**—What has happened? This type of query only presents facts.
- **Diagnostic Query (Past)**—Why did it happen? This type of query presents facts plus an interpretation to provide context.

- **Predictive Query (Future)**—What will happen? This type of query foresees the future according to the facts.
- **Prescriptive Query (Change the Future)**—How to make it happen? What will happen if we do this? This type of query attempts to alter the future.

4.3.2 DDDM Query Type Examples

All four query types are common activities in people's daily lives. Here are intuitive everyday examples:

Describing a scenario in a photograph addresses the *descriptive query* of "*What has happened.*"

Distance traveled and driving speed data presented on an automobile dashboard help explain fuel efficiency levels and address the *diagnostic query* of "*Why did it happen.*"

A GPS navigation system that offers turn-by-turn directions, estimated arrival time, and traffic alerts presents drivers with what to expect on their journey and addresses a *predictive query* of "*What will happen.*"

Asking a personal finance application for guidelines to optimize a budget based on spending habits and financial goals addresses the *prescriptive query* of "*How to make it happen.*"

For product managers, the *prescriptive query* is the most critical as it offers an opportunity for prediction and future manipulation of people's behavior.

4.3.3 Query Format and Components

The query format for product managers is as follows:

> "*As a product manager, I want to know [**question**] because [**objective**], so I need [**data**].*"

Every query has three main components:

- **Question**—This is the information sought, and it pertains to a need, problem, challenge, opportunity, or hypothesis. A hypothesis is a concept or idea to be tested through research and experiments.
- **Objective**—This is the reason why the information is sought.
- **Data**—These are the facts being sought which will answer the question.

4.3.4 Query's Question Component Examples

"As a product manager, I want to know [question]":

- "Does web signup take too long?"
- "Can customer churn be prevented?"
- "Can customer loyalty be improved?"
- "How do we best segment the market?"
- "In which country do most users reside?"
- "What are support ticket response times?
- "Which products our competition is promoting?"
- "Which advertising channel has the best reach?"

4.3.5 Descriptive Query (What Has Happened) Example

*"As a product manager, I want to know [**question**: if web website registrations have declined] because [**objective**: Website registrations are converted to newsletter subscribers, and we use the newsletter to send them product promotions], so I need [**data**: website traffic data]."*

4.3.6 Diagnostic Query (Why Did it Happen) Example

*"As a product manager, I want to know [**question**: if web website registration is taking too long] because [**objective**: there was a decline in website registrations, which we want to encourage for promotional needs], so I need [**data**: website traffic data]."*

4.3.7 Predictive Query (What Will Happen) Example

*"As a product manager, I want to know [**question**: if website registrations will continue to decline] because [**objective**: we want to encourage website registrations for promotional needs], so I need [**data**: an analysis of website traffic data]."*

4.3.8 Prescriptive Query (How to Make it Happen) Example

*"As a product manager, I want to know [**question**: if reducing website registration time by 40% will increase website registrations] because [**objective**: we want to encourage website registrations for promotional needs], so I need [**data**: linear regression of website traffic data]."*

Note: Website traffic data are collected metrics, including the number of visitors, number of users, origin, number of clicks, duration of visits, time spent on the website, bounce rates, and more.

4.3.9 Query Format and Components Retention Drill

1. What are the three main components of a DDDM query?

    (a) Data
    (b) Fact
    (c) Objective
    (d) Reason
    (e) Question
    (f) User

2. *"What has happened?"* Which type of DDDM query is reflected in this statement?

    (a) Predictive Query
    (b) Prescriptive Query
    (c) Diagnostic Query
    (d) Descriptive Query

3. *"Why did it happen?"* Which type of DDDM query is reflected in this statement?

   (a) Prescriptive Query
   (b) Diagnostic Query
   (c) Predictive Query
   (d) Descriptive Query

4. *"What will happen?"* Which type of DDDM query is reflected in this statement?

   (a) Predictive Query
   (b) Prescriptive Query
   (c) Diagnostic Query
   (d) Descriptive query

5. *"How to make it happen?"* Which type of DDDM query is reflected in this statement?

   (a) Descriptive Query
   (b) Predictive Query
   (c) Diagnostic Query
   (d) Prescriptive Query

   Answers: 1-ace, 2-d, 3-b, 4-a, 5-d.

4.4 Crafting Questions

4.4.1 Clear Inquiries

Crafting questions is designing clear, relevant inquiries to guide data exploration.

It is crucial to properly craft the question component of a DDDM query to avoid creating confusing or misleading questions.

When crafting questions for a DDDM query, the focus is on the concept or idea, avoiding the appearance of complexity, bias, leading language, and double negatives.

4.4.2 Complexity Error Example

The following question is undesirably complex and convoluted:

> *"As a product manager, I want to know [**question**: if the cyan-colored interface button's relative subjective value is more or less than the resultant objective proposition of a magenta-colored interface button] because..."*

The following similar question avoids complexity errors:

> *"As a product manager, I want to know [**question**: if cyan is better than magenta as a button's color] because..."*

4.4.3 Bias or Leading Error Example

The following question is biased and leading as it assumes there is a problem with a specific product feature:

> *"As a product manager, I want to know [**question**: what problems do users have with this product feature] because..."*

The following similar question avoids bias or leading error:

> *"As a product manager, I want to know [**question**: are users likely to recommend this product feature] because..."*

4.4.4 Double Negatives Error Example

The following question introduces double negatives, which reduce comprehension and readability:

> *"As a product manager, I want to know [**question**: if it is not untrue that users do not prefer a user interface that is not modern] because..."*

The following similar question avoids double negative errors:

*"As a product manager, I want to know [**question**: if users prefer a classic user interface] because..."*

Optionally, a more controlled and specific question can be:

*"As a product manager, I want to know [**question**: what type of user interface, classic or modern, do users prefer] because..."*

4.5 Metrics Selection

4.5.1 Choosing Indicators

Metrics selection is choosing specific performance indicators or measures for evaluation.

In the DDDM query statement, the data component is reflected in metrics, which consist of variables measured to support the query.

Metrics are sometimes referred to as Key Performance Indicators (KPIs) or Objectives and Key Results (OKRs).

4.5.2 Metric Categories

There are two metrics categories important for product management queries:

- **Adoption**—First-time use of a product feature or product by a user.
- **Engagement**—Ongoing user interaction with a product feature or product.

4.5.3 Adoption Metrics

The Adoption Rate and Time to First Action (TFA) are the adoption metrics categories that are important for product management queries:

- **Adoption Rate**—The percentage of users who interacted with the product or a specific product feature for the first time within a time frame.
- **Time to First Action (TFA)**—Mean time for new users accessing an existing product feature or existing users accessing a new product feature for the first time. Therefore, there are two types of TFA metrics:

  - **New Users Using Existing Product Features**—Mean time for the first use of existing product features by new users.
  - **New Product Features Used by Existing Users**—Mean time for the first use of new product features by existing users.

4.5.4 Adoption Rate Example

During one month, 100 people signed up for the product, from which 40 people became active users.

The adoption rate formula is:

$$(\text{New Active Users} \div \text{Signups}) \times 100 = \text{Product Adoption Rate}$$

In this example, the calculation is:

$$(40 \text{ new active users} \div 100 \text{ signups}) \times 100 = 40\%$$

Accordingly, the product's adoption rate is 40% for that one month.

4.5.5 Adoption Rate TFA Example

Following are examples of Time to First Action (TFA) metric statements:

- **New Users Using Existing Product Features**—Mean time for the first use of new product features by existing users is <u>four calendar days</u>.
- **New Product Features Used by Existing Users**—Mean time for the first use of existing product features by new users is <u>two business days</u>.

4.5.6 Adoption Metrics Retention Drill

1. What do Adoption Metrics measure?

    (a) Ongoing engagement with a product feature.
    (b) Frequency of user interactions with one product feature.
    (c) First-time purchase of a product by a paying customer.
    (d) First-time use of a product feature or product by a user.

2. What do Engagement Metrics measure?

    (a) Ongoing user interaction with a product feature or product.
    (b) Communication effectiveness between product development and product management teams.
    (c) Frequency of product updates a user initiates each month.
    (d) Ongoing buyer interaction with product support.

3. What does the Adoption Rate metric measure?

    (a) The percentage of users who renewed their paid subscription every month.
    (b) The number of users who leave positive product reviews within a time frame.
    (c) The number of users who positively engage in digital content or platforms.
    (d) The percentage of users who interacted with the product or a specific product feature for the first time within a time frame.

4. In the Time to First Action (TFA) adoption metrics category, what does the "New Users Using Existing Product Features" metric measure?

    (a) Mean time for trying out a product's feature set by new users.
    (b) Mean time for new users to complete the onboarding process.
    (c) Mean time for the first use of existing product features by new users.
    (d) Mean time for new users to provide feedback.

5. In the Time to First Action (TFA) adoption metrics category, what does the "New Product Features Used by Existing Users" metric measure?

    (a) Mean time for existing users to submit a support ticket.
    (b) Mean time for the first use of new product features by existing users.
    (c) Mean time for existing users to request new product features.
    (d) Mean time for existing users to use new complex product features.

Answers: 1-d, 2-a, 3-d, 4-c, 5-b.

4.5.7 Engagement Metrics

Engagement metrics are variables that measure ongoing interaction with a product feature or product.

The five most common engagement metrics are:

- **Active Users**—The number of unique and active individuals interacting with the product within a specific time period.
- **Depth of Use**—The quality and richness of user interactions beyond surface-level engagement.
- **Time Spent**—The number of sessions and the duration the users actively spend with the product.
- **Usage Frequency**—How often users return to engage the product, reflecting habitual behavior.
- **Usage Recency**—The elapsed time since the user's last interaction with the product.

4.5.8 Engagement Metrics Example

WhatsApp is an instant messaging and voice-over-IP service that Meta (Facebook) owns.

An example of the five common engagement metrics is found in the 2019 WhatsApp Status Report:

- **Active Users**—500,000 million daily active WhatsApp users worldwide.

- **Depth of Use**—Chat (text and media), P2P voice.
- **Time Spent**—Mean (average) of 5 sessions per day, mean (average) total of 28 minutes per day.
- **Usage Frequency**—58% of active users used WhatsApp more than once per day.
- **Usage Recency**—Mean (average) of 5 hours between engagements.

4.5.9 Engagement Metrics Retention Drill

1. Which of the following best describes Active Users?

   (a) The number of active user accounts currently logged into the system, regardless of their activity level.

   (b) The number of unique and active individuals interacting with the product within a specific time period, accounting for regular engagement and activity.

   (c) The total number of users downloading from all platforms, including initial and repeated downloads.

   (d) The number of customer support tickets submitted by individuals seeking assistance or reporting issues.

2. What does Depth of Use measure?

   (a) The number of product features with comprehensive functionality available to users in the product's feature set.

   (b) The quality and richness of user interactions beyond surface-level engagement, reflecting deep and meaningful usage patterns.

   (c) The length of time users remain registered to the product and their long-term commitment as paid users.

   (d) The frequency of product updates that add new product features to the product.

3. What does Time Spent refer to?

   (a) The total time users spend on the website across all sessions, including both active and idle time.

   (b) The length of time users spend on each product feature that offers insight into product feature popularity and usage.

(c) The average time users need to complete a specific task or activity within the product.

(d) The number of sessions and the duration the users actively spend with the product which indicates the users' engagement and involvement.

4. What does Usage Frequency measure?

   (a) How often users return to engage the product, reflecting habitual behavior and user retention.
   (b) The number of times users log in daily that shows their daily engagement with the product.
   (c) The frequency of manual product updates initiated by the users, highlighting their need for product stability and improvement.
   (d) The number of new users who modify product settings more than once a day.

5. What does Usage Recency indicate?

   (a) The number of users returning daily to use the product, showing daily active user engagement.
   (b) The number of users active since the last week, providing a weekly snapshot of user activity.
   (c) The elapsed time since the user's last interaction with the product, indicating recent activity levels.
   (d) The average time between user sessions, measuring the interval of user engagement.

Answers: 1-b, 2-b, 3-d, 4-a, 5-c.

4.5.10 North Star Metric

The North Star Metric is a product variable the entire company focuses on to achieve long-term growth.

The North Star Metric represents the main business the company is engaged in, highlights the core value of the product, and reflects the user's engagement or activity level.

Because of its prominence, the North Star Metric variable could be used as both the **Objective** and **Data** components in DDDM queries.

North Star Metric categories include consumption, engagement, market share, revenue, and user satisfaction:

- **Consumption**—The extent to which users utilize a product, reflecting their usage intensity.
- **Engagement**—The level of user interaction and activity with the product, often measured by metrics like daily active users or time spent.
- **Market Share**—The portion of a market that a company's product captures relative to competing products.
- **Revenue**—The total money a business earns from selling its products.
- **User Satisfaction**—Level of positivity or negativity that users perceive in their interactions with a product, often measured through Net Promoter Score (NPS) or other feedback mechanisms.

The advantages of having a North Star Metric are that it enables faster decision-making and promotes strategic clarity and focus among all company employees.

However, the disadvantages of the North Star Metric are that it over-optimizes only one business aspect, potentially creates a tendency to overlook other information, and may lead to adverse consequences (Cobra Effect).

4.5.11 North Star Metric Unintended Consequences

An example of undesired business side effects due to an over-focus on one metric is known as the Law of Unintended Consequences or the Cobra Effect.

The Cobra Effect refers to an incentive or policy that results in an opposite target goal.

Many people in India died from Cobra snake bites during the nineteenth century.

To combat this mortal problem by reducing the Cobra snake population, the British Empire initiated a bounty program and paid a fee for every serpent brought in during the British reign over India.

Contrary to its original intent, the bounty program encouraged many Indians to farm Cobra snakes for money, creating an entire industry of growing snakes for income.

Many Cobra snakes were released into the wild when the British terminated the bounty program, ultimately resulting in a dramatic increase in the Cobra snake population in India.

The Cobra Effect can also occur at a company, especially with the North Star Metric.

Unintended consequences occurred when the CEO of an automotive manufacturer set the North Star Metric for the entire company to be the company's stock price, which led to some imaginative accounting and unscrupulous marketing that were eventually exposed, leading to a decline in the company's stock price.

4.5.12 North Star Metric Example (Table 4.1)

Table 4.1 North Star Metric examples

Company	Business	Core Value	North Star Metric
Amazon	eCommerce	Easy Online Shopping	Purchases per subscriber
Facebook	Social Media	Build Community	Time spent actively engaging with feed
Medium	Knowledge Platform	Share Knowledge	Total reading time per user
Netflix	Entertainment	Video On Demand (VOD)	Number of subscribers watching a number of hours
Salesforce	Relationship Management	Customer Engagement	Number of records per account
WhatsApp	Messaging	Immediate Interaction	Messages sent per user

4.5.13 North Star Metric Retention Drill

1. Which of the following best describes the North Star Metric in a business context?

    (a) Financial indicator used to measure short-term revenue performance.
    (b) Key performance indicator (KPI) that evaluates employee satisfaction and engagement.
    (c) Marketing metric for measuring the effectiveness of advertising campaigns.
    (d) Product variable the entire company focuses on to achieve long-term growth.

2. The North Star Metric variable could be used as both the _____ and _____ components in DDDM queries.

    (a) Questions
    (b) Objective
    (c) User
    (d) Data

3. Which two categories do not determine a North Star Metric?

    (a) Consumption
    (b) Engagement
    (c) Development Speed
    (d) Market Share
    (e) Revenue
    (f) Acquisition
    (g) User Satisfaction

4. Which two are the advantages of the North Star Metric?

    (a) Enables faster decision-making.
    (b) Reduces the need for market research.
    (c) Enhances customer satisfaction.
    (d) Promotes strategic clarity and focus.

5. What is a potential disadvantage of using the North Star Metric?

    (a) Overemphasize one business metric at the expense of other metrics.
    (b) Encourages metric alignment and focus among all company departments.
    (c) Discourages data reliance and intuition when formulating strategies.
    (d) Complicates the decision-making process by requiring complex analysis.

6. What does the term "Cobra Effect" refer to in a business context?

    (a) An economic strategy that maximizes profit through market manipulation.
    (b) An incentive or policy that results in an opposite target goal.
    (c) A risk management approach that mitigates potential losses.
    (d) A marketing campaign designed to increase brand loyalty.

Answers: 1-d, 2-bd, 3-cf, 4-ad, 5-a, 6-b.

4.5.14 AARRR Framework

In 2007, Dave McClure, venture capitalist, angel investor, and founder of the "500 Startups" startup accelerator, introduced a 5-step framework for growth.

The framework was called AARRR, or the Pirate Metrics.

The AARRR framework is meant to help select the appropriate metrics for growth and is intended for startups and digital products.

Like the North Star Metric, metrics within the AARRR framework could be used in the **Data** component in DDDM queries.

The AARRR acronym stands for acquisition, activation, retention, referral, and revenue:

- **Acquisition**—Understanding where users come from, including search, social media, marketing campaigns, apps, referrals, advertising, or organic discovery.

- **Activation**—Assessing the quality of a user's initial experience with the product. Ensuring a positive initial interaction is crucial and accomplished with quick signup processes, intuitive interfaces, and effective onboarding.
- **Retention**—Tracking how many users are retained over time and understanding why some leave. The goal is to build loyalty and reduce churn.
- **Referral**—Turning existing users into advocates who recommend the product to others through referral programs, incentives, and social sharing product features.
- **Revenue**—Devising strategies to increase overall revenue and monetize the user base through pricing optimization, upsells, or expanding the user base.

4.5.15 AARRR Framework Retention Drill

1. What does the AARRR acronym stand for?

    (a) Attract, activate, retain, recommend, and revenue.
    (b) Attention, attraction, relationship, recommendation, and returns.
    (c) Awareness, assessment, resolution, re-engagement, and results.
    (d) Acquisition, activation, retention, referral, and revenue.

2. Which of the following best describes the "Acquisition" concept in the AARRR framework?

    (a) Identifying the usability metrics through which users engage the product.
    (b) Analyzing the effectiveness of user onboarding processes.
    (c) Understanding where users come from, such as marketing campaigns.
    (d) Evaluating the initial user interactions with the product.

3. What does "Activation" focus on in the AARRR framework?

    (a) Guaranteeing users receive prompt customer support and resolution.
    (b) Ensuring users have a positive initial experience with the product.

(c) Providing users with immediate access to premium features.
(d) Making sure users complete their first purchase quickly.

4. What is the "Retention" idea in the AARRR framework concerned with?

   (a) Attracting new users through marketing campaigns.
   (b) Improving the initial user experience to encourage signups.
   (c) Building loyalty and reducing churn over time.
   (d) Encouraging existing users to refer new customers.

5. What does "Referral" in the AARRR framework aim to do?

   (a) Turning users into advocates through incentives and social sharing.
   (b) Encouraging users to provide general feedback and product suggestions.
   (c) Attracting new users through targeted marketing campaigns and ads.
   (d) Building user loyalty through rewards and exclusive offers.

6. What does the "Revenue" element of the AARRR framework focus on?

   (a) Developing premium subscription models and pricing plans.
   (b) Increasing overall revenue and monetizing the user base.
   (c) Implementing upsell opportunities and cross-sell strategies.
   (d) Expanding revenue streams through partnerships and sponsorships.

Answers: 1-d, 2-c, 3-b, 4-c, 5-a, 6-b.

4.5.16 Typical AARRR Metrics

Acquisition Metrics:

- **Bounce Rates**—The percentage of users who hastily leave the product with minimal or no interaction with the product. This metric indicates user engagement levels and provides insights into how effective a website's content and user experience are.

- **Click-Through Rate (CTR)**—Percentage of users engaging with the product's call-to-action feature. This metric indicates how well an advertisement or website link captures the audience's interest and prompts them to take action.
- **Conversion Rate**—The percentage of users who perform any desired action. This metric indicates the effectiveness of a marketing campaign, website, or any other initiative intended to encourage a specific action from users.
- **Cost Per Click (CPC)**—The price paid for each interaction in online advertising campaigns. This metric indicates the cost-efficiency of online advertising campaigns.
- **Customer Acquisition Cost (CAC)**—The expenditure to attract new customers through various channels. This metric indicates how effectively a company spends its marketing and sales budget to acquire new customers.
- **Dwell Time**—The duration of the user's session with the product. This metric indicates content effectiveness and quality of the user experience.
- **Traffic Source**—Tracking and listing the origin and ratio from where users came to the product. This metric indicates how users find their way to their digital platforms and which channels are most effective in driving traffic.

Activation Metrics:

- **Conversion Rate**—The percentage of users who perform a specific action, such as registering. This metric indicates marketing effectiveness or all effort types to get users to complete desired actions.
- **Time to Value**—The mean time for users to recognize the product's benefits. This metric indicates onboarding effectiveness, product usability, and user satisfaction levels.
- **Visitors to Registration Ratio**—The percentage of users registering for the product. Also, specifically in response to a marketing effort. This metric indicates the effectiveness of a marketing campaign in converting visitors to registered users.

Retention Metrics:

- **Churn Rate**—The percentage of registered users who entirely cease using the product. This metric indicates the rate at which companies lose customers, reflecting on demand for the product and the overall health and growth of a business.
- **Retention Rate**—The percentage of users who continue using the product over time. This metric is the opposite of the churn rate metric and provides insights into customer loyalty and the effectiveness of a business in keeping its users engaged.

Referral Metrics:

- **Net Promoter Score (NPS)**—The probability that users will recommend the product to others. This metric indicates customer satisfaction and loyalty, signals good standing with users, and alerts to potential issues.
- **Purchase Rate of Referred Customers**—The percentage of referred users that become paying customers. This metric indicates the effectiveness of a referral program in converting referred leads into paying customers.
- **Referral Rate**—The percentage of existing users who refer others to the product. This metric indicates how effective and engaged the current user base is in promoting the product independently or through referral programs.
- **Viral Cycle Time**—The mean time it takes a user to refer others to the product. This metric indicates how quickly users let others know about the product.

Revenue Metrics:

- **Average Revenue per User (ARPU)**—The average revenue generated from each user or customer within a specific period, calculated by dividing the total revenue from all users by the total number of users. This metric indicates how much money each user contributes to a business.

- **Customer Lifetime Value (CLV)**—The predicted total amount of money a user will bring to the business over their entire engagement with the company. This metric indicates the long-term value of acquiring and retaining customers.
- **Revenue Growth Rate (RGR)**—The increase in a company's total revenue over a specified period, typically expressed as a percentage. This metric indicates the company's ability to increase sales, expand its market share, and improve its financial health over time.

4.5.17 Metadata

Metadata is data that describes other data, but it is not the other data itself.

For the most part, and relative to product management, metadata is of lesser interest, but in certain cases, metadata could be used in the **Data** component in DDDM queries.

Product managers decide if to track, collect, and analyze metadata.

4.5.18 Metadata Examples

The following are examples of data and related metadata.

The data is the content in a document, and the metadata are the document's author, file size, create date, modified date, and keywords.

The data is the music score, and the metadata are the composer, singer, album, length, and release date.

The data is a phone conversation's content, what people talked about, and the metadata are the call's time, date, duration, caller and recipient names, phone numbers, and location.

The data is a web page's content, and the metadata are the webpage's description, tags, charset, and author.

4.6 Data Inventory

4.6.1 Organizing Data

Data inventory is a systematic process within DDDM that involves cataloging and organizing all data assets by identifying, documenting, and classifying data sources, types, formats, and storage locations.

The primary aim of a data inventory is to ensure data accessibility, accuracy, and compliance while providing a comprehensive overview of available data resources to support informed decision-making and strategic planning.

To aid in completing the **Data** component in DDDM queries, data inventory checks for the following:

- Data that is needed.
- Data the company presently has.
- Data the company does not have and should collect or purchase.
- Gaps in the company's data (how complete is the data).

The data held by the company can be categorized as either Primary Data, which is data collected directly by the company itself, or Secondary Data, which is data previously collected by other parties.

4.6.2 Open Data Concept

Open Data refers to freely available data for the public to use and share without restrictions from copyright, patents, or regulatory constraints.

The primary goal of open data is to promote transparency, foster innovation, and enable informed decision-making by making valuable data accessible to all.

Open data can come from various sources, including government agencies, research institutions, and private companies, and often includes information on topics such as economic indicators, environmental conditions, and public services.

The concept of open data gained traction in the early 2000s, with significant milestones including releasing government datasets and establishing open data initiatives.

A notable early example is the U.S. government's launch of the data.gov website in 2009, which aimed to provide public access to federal datasets.

The government's open data website contains free data, tools, and resources to conduct research, develop web and mobile applications, and design.

The open data movement has grown globally, with many countries and companies adopting open data policies to enhance transparency and civic engagement.

4.7   Data Collection

4.7.1   Gathering Data

Data collection is the systematic process of gathering data from different sources to answer specific research questions, test hypotheses, and evaluate outcomes.

Data collection sources include:

- Interviews
- Observations
- Focus groups
- Questionnaires
- Online tracking

4.7.2   Interviews

Interviews are a qualitative data collection method involving direct interaction between an interviewer and interviewee to gather detailed information on thoughts, experiences, and perceptions.

Interviews can be structured, semi-structured, or unstructured, allowing for in-depth exploration of complex issues.

Interviews, whether in-person or over the phone, are customizable and responsive, allowing for open-ended questions, but they are also the most expensive and time-consuming method.

4.7.3 Observations

Observations involve systematically watching and recording behaviors, events, or conditions in their natural setting.

Observations provide real-time data and are useful for studying interactions and behaviors that are difficult to capture through other methods.

Observations involve watching an individual's behavior and collecting data without asking questions, but they are considered subjective and interpretative.

4.7.4 Focus Groups

Focus groups are a qualitative data collection method where a trained moderator leads a guided group discussion to gather insights on participants' perceptions, opinions, and attitudes towards a specific topic.

Typically composed of 6–12 participants, focus groups facilitate dynamic interaction and exploration of diverse perspectives.

Focus groups, which involve interviewing and observing a group of individuals, add a collective element to the data set, but the results can be biased and affected by group dynamics.

4.7.5 Questionnaires

Questionnaires are a data collection tool consisting of questions designed to gather information from respondents.

Questionnaires can include closed-ended questions for quantitative data and open-ended questions for qualitative insights, administered in various formats such as paper-based, online, or mobile.

Questionnaires are good for closed-ended questions and are effective for gathering large amounts of data from a large population, but they can be prone to bias.

4.7.6 Online Tracking

Online tracking constantly collects and stores user activity data with a digital product.

Online tracking involves collecting data on users' behaviors and interactions within digital environments using technologies like web analytics, cookies, and tracking pixels.

Online tracking provides quantitative data on metrics such as page views, clicks, user pathways (how users move through a website or application), and qualitative insights into user interaction patterns.

Online tracking collects enormous data sets, generates accurate real-time data, and is directly tied to data tracking software.

4.7.7 Online Tracking Specification

Online tracking is crucial for product management because it offers valuable insights into how users interact with a product, thus enabling data-driven decision-making.

Online tracking is the most common way of gathering data to make product decisions for digital products.

Collecting actionable data from user interactions with your product begins with preparing an **Online Tracking Specification** used by analytics software that tracks user interactions with web and mobile applications.

The **Online Tracking Specification** lists the user interactions and associated data points to be tracked.

As a product manager, this step is the most critical to do correctly, as mistakes or omissions at this stage can result in collecting the wrong data, making it difficult to derive meaningful insights.

Based on the nomenclature used in popular tracking software systems, the **Online Tracking Specification** consists of three components:

1. Event
2. Event Properties
3. User Properties

4.7.8 Event

An event is a single user interaction with a digital product that is being tracked.

Examples of events in a website include clicks, hovers, scrolls, touches, app openings, and mouse cursor positions.

4.7.9 Event Properties

Event properties are additional data points for the user interaction event.
Event properties capture the context around an event.
Event properties in a website include page URL, time, date, browser, and device.

4.7.10 User Properties

User properties capture the context around the user performing the event.
Examples of user properties related to a website event include the user's gender, age, location, and user category (registered or unregistered, paid or free).

4.7.11 Amazon Online Tracking Example

A typical Amazon **Online Tracking Specification** is:

1. **Event**—Add to Cart button is clicked.
2. **Event Properties**—Product, browser, device, date/time.
3. **User Properties**—Gender, location, visit frequency.

The outcome of implementing an **Online Tracking Specification** is an **Online Tracking Record**:

"A male user from Switzerland, who visits Amazon's European website twice a week and uses a Safari browser on an iPhone 7, added a Bosch hammer drill (product code 26497586) to the shopping cart, on Friday, June 26, 2020, at 5:32 pm."

Amazon utilizes data to determine which products to recommend to customers based on the users' previous purchases and search behavior patterns.

Amazon leverages data analytics and machine learning to power its recommendation engine.

It is estimated that over 35% of Amazon's consumer purchases were attributable to Amazon's recommendation system.

4.7.12 Data Collection Retention Drill

1. Regarding data collection, what are interviews?

    (a) Qualitative method using online interactions for comprehensive data collection.
    (b) Qualitative method involving group discussions for varied insights.
    (c) Qualitative method using indirect questions to gather subtle information.
    (d) Qualitative method involving direct interaction to gather detailed information.

2. Regarding data collection, what are observations?

    (a) Systematically watching and recording behaviors, events, or conditions in their natural setting.
    (b) Systematically recording reactions and interactions in a controlled environment.
    (c) Systematically monitoring and documenting responses and actions in a simulated setting.
    (d) Systematically observing and noting behaviors and events in an experimental context.

3. Regarding data collection, which of the following best describes focus groups?

   (a) Quantitative method where individual interviews are conducted to collect numerical data.
   (b) Qualitative method where participants fill out surveys to provide their opinions.
   (c) Qualitative method where a moderator leads a guided group discussion for insights.
   (d) Quantitative method where data is collected through structured interactions.

4. Regarding data collection, what are questionnaires?

   (a) Method of analyzing data and generating insights from existing information.
   (b) Data collection tool consisting of questions designed to gather information from respondents.
   (c) Process for interviewing respondents to obtain detailed personal stories.
   (d) Technique for observing and recording behaviors in a natural setting.

5. Regarding data collection, which of the following best describes online tracking?

   (a) Online tracking intermittently gathers user preferences and habits for targeted advertising.
   (b) Online tracking constantly collects and stores user activity data with a digital product.
   (c) Online tracking occasionally compiles user feedback to improve digital interfaces.
   (d) Online tracking regularly surveys user opinions and satisfaction levels for product development.

6. Regarding data collection, what are the three common components of an "Online Tracking Specification"?

   (a) Event
   (b) Design
   (c) Event Properties
   (d) Design Properties
   (e) Protocol
   (f) User Properties

Answers: 1-d, 2-a, 3-c, 4-b, 5-b, 6-acf.

4.8 Data Preparation

4.8.1 Data Cleaning

Data preparation, also known as data cleaning, involves the meticulous and systematic transformation of raw data into a format suitable for analysis by removing or correcting incorrect, incomplete, or irrelevant data.

This process encompasses several key tasks, including identifying and handling missing data, correcting data entry errors, removing duplicate records, and normalizing and standardizing data to ensure consistency.

The primary objective of data preparation is to improve data quality by addressing inaccuracies and inconsistencies, thereby ensuring that the data is reliable and valid for subsequent analytical tasks.

Data preparation includes the following core tasks:

- **Data Validation**—Checking for and rectifying errors such as invalid entries, outliers, and discrepancies.
- **Data Transformation**—Converting data into a consistent format, which may include standardizing units of measurement, normalizing text data, and encoding categorical variables.
- **Data Integration**—Combining data from multiple sources, ensuring that the integrated data is coherent and free from redundancy.

- **Data Imputation**—Replacing missing data values with substituted values using mean, regression, or multiple imputations.

Effective data preparation enhances the accuracy and efficiency of data analysis, leading to more robust and reliable insights. It is an essential step that underpins the integrity of any data-driven decision-making process.

Data preparation is done by data professionals, such as **Data Analysts**, who often spend 80% of their time collecting, cleaning, and organizing data and only 20% performing data analysis.

4.8.2 Data Preparation Retention Drill

1. Which four core tasks comprise "Data Preparation"?
    (a) Data Validation
    (b) Data Analysis
    (c) Data Exploration
    (d) Data Transformation
    (e) Data Mining
    (f) Data Science
    (g) Data Integration
    (h) Data Imputation

2. What does Data Validation involve?
    (a) Checking for and fixing errors like invalid entries and outliers.
    (b) Identifying and adding data values to incomplete data sets.
    (c) Ensuring all data entries follow a consistent numerical format.
    (d) Verifying and correcting missing data points.

3. What is the focus of Data Transformation?
    (a) Ensuring data entries are error-free.
    (b) Integrating multiple datasets into a unified dataset.
    (c) Replacing missing data with substituted values.
    (d) Converting data into a consistent format.

4. What is the purpose of Data Integration?

   (a) Ensuring all data entries follow a consistent format.
   (b) Replacing missing data values with estimated ones.
   (c) Combining data from multiple sources for coherency.
   (d) Transforming data into a standardized structure.

5. What does Data Imputation entail?

   (a) Using computers to convert data into a consistent format.
   (b) Substituting missing values with mean or regression.
   (c) Checking for and fixing errors like invalid entries.
   (d) Replacing missing data values with virtual values.

Answers: 1-adgh, 2-a, 3-d, 4-c, 5-b.

4.9 Data Analysis

4.9.1 Drawing Insights

Data analysis systematically examines, transforms, and arranges data sets to study and extract valuable insights.

Data analysis is fundamental in understanding data structures and deriving actionable information to support decision-making.

Data analysis includes the following core tasks:

- **Statistical Calculations**—Applying statistical methods to analyze data distributions, relationships, and variations. Data analysts can summarize data characteristics and infer patterns or trends through statistical calculations.
- **Pattern Recognition**—Identifying recurring patterns or regularities within the data set is essential for making predictions and identifying significant trends. Pattern recognition techniques enable the detection of meaningful structures within the data.

- **Detecting Consistencies and Inconsistencies**—Ensuring data integrity is crucial for accurate analysis. This task involves identifying and resolving discrepancies or anomalies in the data to maintain consistency and reliability.

Data analysis transforms raw data into meaningful information, facilitating informed decision-making and strategic planning.

4.9.2 Data Analysis Retention Drill

1. What are the three core tasks in "Data Analysis"?
   (a) Data Modeling
   (b) Statistical Calculations
   (c) Pattern Recognition
   (d) Machine Learning
   (e) Detecting Consistencies and Inconsistencies
   (f) Statistical Visualization

2. Data analysis transforms raw data into meaningful _____.
   (a) Numbers
   (b) Statistics
   (c) Information
   (d) Actions

   Answers: 1-bce, 2-c.

4.10 Data Visualization

4.10.1 Graphic Format

Computers are inherently designed to process numerical data, efficiently navigating through extensive rows and columns of numbers.
  In contrast, humans are more attuned to visuals, finding it easier to comprehend data when presented in a meaningful, visual format.

Data dashboards, also called data visualization software, are software applications that graphically represent data and information in real time or as part of an analysis.

Data dashboards transform textual and numerical data into visual charts, figures, and tables, making the data and the resulting information more accessible and understandable for users.

Often, data dashboards serve as tools for application or performance monitoring by graphically presenting, potentially in real-time, crucial data on a central interface while being continuously linked to the data sources.

There are three types of data dashboards: operational dashboards, strategic dashboards, and analytical dashboards.

- **Operational Dashboards**—These dashboards display key metrics in real-time, providing immediate insights into current performance and allowing for prompt decision-making and action.
- **Strategic Dashboards**—These dashboards summarize key metrics over a specified period, offering a comprehensive overview that supports long-term planning and strategic decision-making.
- **Analytical Dashboards**—These dashboards delve deeper into data, analyzing and presenting trends, predictions, and insights, enabling more informed and predictive decision-making processes.

4.10.2   Data Visualization Retention Drill

1. Humans find it easier to comprehend data when presented in which format?
    (a) Numerical
    (b) Virtual
    (c) Organized
    (d) Visual

2. What is the name of software applications that graphically represent data and information in real time or as part of an analysis?

   (a) Data Applications
   (b) Data Dashboards
   (c) Virtualization Software
   (d) Imaging Systems

3. What are the three types of data dashboards?

   (a) Software Dashboards
   (b) Operational Dashboards
   (c) Tactical Dashboards
   (d) Strategic Dashboards
   (e) System Dashboards
   (f) Analytical Dashboards

4. Which type of data dashboards display key metrics in real-time, providing immediate insights into current performance and allowing for prompt decision-making and action?

   (a) Analytical Dashboards
   (b) Strategic Dashboards
   (c) Operational Dashboards
   (d) Software Dashboards
   (e) Tactical Dashboards
   (f) System Dashboards

5. Which type of data dashboards summarize key metrics over a specified period, offering a comprehensive overview that supports long-term planning and strategic decision-making?

   (a) Operational Dashboards
   (b) Analytical Dashboards
   (c) Strategic Dashboards
   (d) Software Dashboards
   (e) Tactical Dashboards
   (f) System Dashboards

6. Which type of data dashboards delve deeper into data, analyzing and presenting trends, predictions, and insights, enabling more informed and predictive decision-making processes?

(a) Analytical Dashboards
(b) Operational Dashboards
(c) Strategic Dashboards
(d) Software Dashboards
(e) Tactical Dashboards
(f) System Dashboards

Answers: 1-d, 2-b, 3-bdf, 4-c, 5-c, 6-a.

4.11 Information Visualization

Information visualization translates information into visual formats such as graphs, diagrams, tables, images, and charts.

Information visualization depicts different types of information that support decision-making: trends, comparisons, relationships, and revelation data.

4.11.1 Trends

A trend is a general direction where something is changing or developing.

Whether rising or falling, trends form the core narrative of many data-driven stories.

For example, in recent years, software companies have significantly increased their investment in privacy compliance tools, driven by regulatory demands and consumer awareness about data privacy.

Surveys have demonstrated that over 85% of users want more privacy and control over their data, which has prompted technology companies to make privacy components a standard product feature (Fig. 4.1).

The graph above demonstrates technology companies' growing investment in privacy compliance tools.

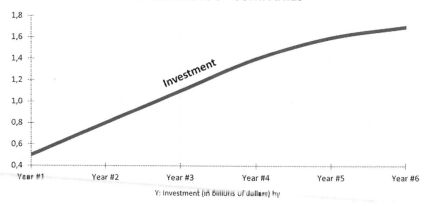

Fig. 4.1 Investment in privacy compliance tools

This trend reflects the industry's response to stricter regulatory requirements and heightened consumer demand for data privacy.

The continuous rise in investment in privacy compliance tools underscores the importance of safeguarding personal data and maintaining trust in the digital age.

4.11.2 Comparisons

Comparisons involve measuring the similarity or dissimilarity between two entities.

A common data-driven narrative is to compare different items.

For example, comparing the growth of active users on x.com (formally Twitter) with Facebook is often done with charts highlighting how Facebook outperforms Twitter in terms of active users' growth (Fig. 4.2).

The graph above demonstrates the growth of active users on X.com (formerly Twitter) and Facebook.

This comparison highlights Facebook's significantly larger user base and continued expansion compared to X.com.

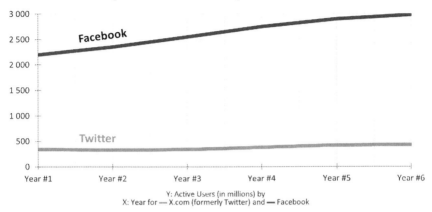

Fig. 4.2 Growth of active users

4.11.3 Relationships

A relationship refers to a correlation or causation between variables.

For example, there is a negative relationship between a disorganized desk and work productivity (Fig. 4.3).

This graph demonstrates the negative relationship between a disorganized desk and work productivity.

As the level of desk disorganization increases from very organized to very disorganized, work productivity decreases significantly.

This visual representation highlights how maintaining an organized workspace can positively impact productivity.

4.11.4 Revelation Data

Revelation involves disclosing previously unknown facts.

For example, Coca-Cola has an extensive global reach and is present in almost every country except North Korea and Cuba, which have long-term U.S. trade embargoes.

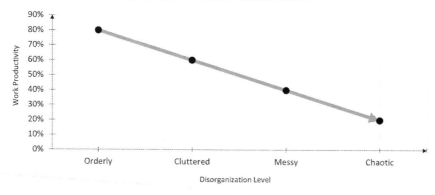

Fig. 4.3 Negative relationship

Interestingly, North Korea produced a dark-colored soda, Ryongjin Cola, mimicking American Coca-Cola's red packaging and cursive font (Fig. 4.4).

The pie chart above illustrates the global presence of Coca-Cola, showing that it is available in nearly every country (195 countries) except North Korea and Cuba.

4.11.5 Effective Information Visualization

The following are best practices for effective information visualization:

- **Clarity**—Visuals that provide relatively simple and easy-to-understand and remember.
- **Support**—Visuals that are directly linked to the decision-making.
- **Focus**—Visuals that are focused on summarizing the information.
- **Key Statistic**—Visuals that highlight one key statistic to emphasize the main point.
- **One Clear Message**—Visuals that convey a single, clear message or insight.
- **Suitable Visuals**—The appropriate visual format is selected for the target audience.

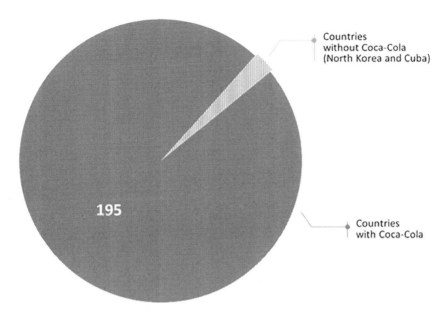

Fig. 4.4 Revelation data

4.11.6 Information Visualization Retention Drill

1. What four types of information does information visualization depict to support decision-making?

    (a) Trends
    (b) Distributions
    (c) Comparisons
    (d) Hierarchies
    (e) Anomalies
    (f) Relationships
    (g) Revelation Data
    (h) Geospatial Data

2. Regarding information visualization, what term best describes the general direction where something changes or develops?

   (a) Comparison
   (b) Relationship
   (c) Revelation Data
   (d) Distribution
   (e) Hierarchy
   (f) Trend
   (g) Anomaly

3. Regarding information visualization, what term best describes measuring the similarity or dissimilarity between two entities?

   (a) Trend
   (b) Relationship
   (c) Revelation Data
   (d) Comparison
   (e) Distribution
   (f) Hierarchy
   (g) Anomaly

4. Regarding information visualization, what term best describes a correlation or causation between variables?

   (a) Trend
   (b) Comparison
   (c) Revelation Data
   (d) Distribution
   (e) Hierarchy
   (f) Anomaly
   (g) Relationship

5. Regarding information visualization, what term best describes disclosing previously unknown facts?

   (a) Trend
   (b) Comparison
   (c) Relationship
   (d) Revelation Data

(e) Distribution
(f) Hierarchy
(g) Anomaly

Answers: 1-acfg, 2-f, 3-d, 4-g, 5-d.

4.12 Communicating Information

In the data-driven decision-making (DDDM) process, the communicating information step is critical to ensure that findings are effectively conveyed to stakeholders.

This step involves the strategic presentation of data insights, analysis results, and recommendations in a clear and comprehensible manner.

Effective communication in this context requires tailoring the presentation to the audience's expertise and interest, ensuring that complex data is translated into actionable insights.

Visual aids such as images, tables, charts, graphs, and dashboards are often employed to enhance understanding and highlight key findings. The goal is to provide stakeholders with a comprehensive yet digestible overview of the data, facilitating informed decision-making.

Engaging stakeholders and eliciting their input during this step is equally important.

Open dialogues and feedback mechanisms are encouraged to ensure stakeholders' perspectives and expertise are considered in decision-making.

A collaborative approach promotes ownership and alignment among stakeholders and enhances the credibility and acceptance of data-driven decisions.

4.13 Making Decisions

Decisions are judgments or choices that instigate action based on acquired knowledge.

In DDDM, making decisions means making informed choices that are guided by data-driven insights.

The value of data-driven decisions hinges on the quality of the data, its analysis, and interpretation.

The essence of the DDDM process for product management is query formulation and product decisions.

Therefore, it is incumbent upon product managers to provide data analysts with clear and properly formed queries that guide data analysts in configuring data dashboards and performing data analysis and data visualization, ultimately yielding the predictions and insights that product managers require making product decisions.

4.13.1 Making Decisions Retention Drill

1. What two activities exemplify the essence of the DDDM process for product management?
   - (a) Query Formulation
   - (b) Statistical Analysis
   - (c) Data Collection
   - (d) Product Decisions
   - (e) Information Visualization

   Answers: 1-ad.

5

Peers and Environment

5.1 The Data Analyst

The data analyst is a technical role that requires proficiency in various analytical tools, statistical methods, and data visualization techniques.

The data analyst's job responsibilities require using software and statistically capable programming languages to manipulate, analyze, and interpret complex data sets.

Data analysts collect, prepare, and statistically analyze large datasets to derive actionable insights.

The data analyst's core duties include data cleaning, mining, visualization, and applying various statistical techniques to interpret complex data sets.

Data analysts utilize tools and languages like SQL, Python, Excel, and various data analysis and visualization platforms, including Graphpad, SPSS, XLSTAT, Minitab, MATLAB, and R software + RStudio.

Data analysts also generate reports, create dashboards, and present their findings to product managers and other internal stakeholders to support data-driven decision-making within a company.

The core responsibilities of a data analyst include:

- **Data Collection and Preparation**—Gathering data from multiple sources and ensuring data integrity and accuracy through cleaning and preprocessing.
- **Data Analysis and Interpretation**—Using statistical methods to identify trends, patterns, and correlations; and interpreting data to provide meaningful insights.
- **Data Visualization**—Creating visual representations of data to communicate findings effectively and developing dashboards and reports accessible to non-technical stakeholders.
- **Collaboration with Stakeholders**—Working with various departments to understand their data needs and providing data-driven recommendations to support business objectives.

Data analysts are critical in the data-driven decision-making (DDDM) process concerning product management.

Data analysts directly support product managers in making informed decisions regarding market research, product decisions, product features, and product strategy.

The collaboration between data analysts and product managers is in the following activities:

- **Market and User Research**—Data analysts provide insights into market trends and user behavior through detailed analysis. These insights help product managers identify opportunities and gaps in the market.
- **Product Planning**—Data analysts help product managers understand how existing products are used by analyzing user feedback and product usage data. This information is crucial for prioritizing product features.
- **Performance Monitoring**—Data analysts track key performance indicators (KPIs) to evaluate product performance. Product managers leverage user engagement, retention rates, and other performance data in DDDM.
- **Strategic Decision-making**—Data analysts offer insights into future market movements through predictive analytics and trend analysis. Product managers rely on these insights to develop long-term product strategies and roadmaps.

- **A/B Testing and Experimentation**—Data analysts design and analyze A/B tests to determine the impact of product changes. The test results guide product managers in making evidence-based decisions on product iterations.

A data analyst's role is integral to product management's success.

Data analysts possess solid mathematical and statistical skills, business acumen, and computer science and coding proficiency.

Data analysts develop key performance indicators (KPIs), create visual representations of data, and utilize business intelligence and analytics tools to derive actionable insights that support product decisions.

By converting raw data into meaningful insights, data analysts empower product managers to make data-driven decisions, optimize product features, and ultimately enhance the overall user experience.

The collaboration between data analysts and product managers ensures that product strategies are backed by solid evidence, fostering a more efficient and effective product development process.

The growing reliance on data and the realization of its complexities have led companies to train product managers in data literacy and hire data analysts.

In collaboration with data analysts, data-literate product managers can leverage data more effectively to drive product innovation and strategic decision-making.

The synergy between data-literate product managers working with data analysts is essential for navigating the increasingly data-driven landscape of modern business.

5.1.1 Data Analyst Retention Drill

1. What type of role does a data analyst have?
   - (a) Operational
   - (b) Technical
   - (c) Strategic
   - (d) Tactical

2. What are the four core responsibilities of a data analyst?

   (a) Software Development
   (b) System Administration
   (c) Data Collection and Preparation
   (d) Project Management
   (e) Data Analysis and Interpretation
   (f) Data Visualization
   (g) Database Administration
   (h) Collaboration with Stakeholders

3. Which is not an area of collaboration between data analysts and product managers?

   (a) Market and User Research
   (b) Product Planning
   (c) Performance Monitoring
   (d) Database Management
   (e) Strategic Decision-making
   (f) A/B Testing and Experimentation

   Answers: 1-b, 2-cefh, 3-d.

5.1.2 Data-driven Culture

A company's data culture refers to the collective mindset and behaviors within a company that prioritizes the use of data in decision-making processes.

Data culture encompasses the values, practices, and standards that guide how data is collected, managed, and utilized across all company levels.

A strong data culture ensures that data is treated as a valuable asset, with clear policies and governance structures to maintain its quality, security, and accessibility.

In a data culture, employees at all levels are encouraged and equipped to leverage data daily, fostering an environment where data-driven insights are integral to strategic planning and operational execution.

Company executives highlight the importance of data literacy, providing the appropriate training and resources to enable product teams to interpret and use data effectively.

A mature data culture also emphasizes collaboration, where data sharing and transparency are standard practices, breaking down silos and enabling cross-functional teams to work more cohesively.

The company continuously invests in advanced analytics tools and technologies to enhance data capabilities, driving innovation and competitive advantage.

A robust company data culture aligns with the company's goals and objectives, ensuring that data-informed decisions lead to improved outcomes and sustained success.

5.1.3 Attitudes Toward Data

Within a company, attitudes toward data can vary significantly.

Some employees exhibit **Data Denial**, where there is a distrust of data, leading to deliberately avoiding its use.

Others are **Data Indifferent**, perceiving no need for data and displaying a general lack of concern about its relevance.

On the other hand, some are **Data Informed**, relying primarily on intuition but using data to support and validate their instincts.

At the most advanced level, a **Data-driven** attitude prevails, where data fundamentally shapes and drives decision-making processes across the company.

Recapping the different attitudes toward data in a company:

- **Data Denial**—Employees with a data denial attitude exhibit distrust towards data and actively avoid its use in their decision-making processes. These employees may question the accuracy or relevance of data, relying instead on subjective judgment or experience.
- **Data Indifferent**—Data indifferent employees lack interest or concern for data. These employees do not perceive the need for data in their roles and are generally disengaged from data-driven practices,

relying on traditional methods and intuition without seeking data validation.
- **Data Informed**—Employees who are data informed use data to support their intuition and decision-making. While these employees value and incorporate data into their processes, their decisions are primarily guided by experience and intuition, with data as a supplementary resource.
- **Data-driven**—Employees with a data-driven attitude rely heavily on data to shape their decisions. These employees prioritize data analysis and insights, ensuring their actions and strategies are grounded in objective, measurable information. This approach emphasizes the importance of data as the foundation for all business decisions.

5.1.4 Building a Data-driven Culture

Building a corporate data-driven culture follows six essential steps:

1. **Executive Support and Example**—A data-driven culture begins at the top. Executives demonstrate their commitment to data-driven decision-making by consistently using data in their strategic and operational decisions. Executives' behavior sets a standard for the rest of the company, emphasizing the importance of data at all levels.
2. **Quantifying Almost Everything**—It is essential to quantify as many business aspects as possible to foster a data-driven culture. Quantification means collecting data on various activities, processes, and outcomes, ensuring that decisions are based on measurable and objective information rather than intuition or anecdotal evidence.
3. **Selecting Simple Metrics**—Choosing simple, clear, and relevant metrics is crucial. These metrics should be easy to understand and communicate, clearly showing performance and progress. Simple metrics help maintain focus on key business objectives and make it easier for employees to align their efforts with company goals.
4. **Easy Access to Information**—Ensuring that data and information are easily accessible to all employees is vital. Creating easy access to data means creating systems and processes that allow for seamless data

flow across the company. Easy access to information empowers employees to make informed decisions quickly and efficiently.
5. **Visualizing Information**—Visualizing data through charts, graphs, and dashboards makes complex information more comprehensible and actionable. Effective data visualization helps identify trends, patterns, and anomalies, enabling better insights and fostering a deeper understanding of the data.
6. **Developing Data Literacy**—Building a data-driven culture requires developing data literacy across the company. Archiving data literacy entails training employees to understand, interpret, and use data effectively. By enhancing data literacy, companies ensure that employees at all levels can contribute to and benefit from data-driven decision-making.

These six steps help create a robust data-driven culture, enabling companies to leverage data for improved decision-making, enhanced performance, and sustained competitive advantage.

Make information immediately accessible and visual to every company employee, and look everywhere for validated data-supported patterns.

5.1.5 Netflix Data-driven Culture Example

Netflix has long utilized data to provide insights that guide its movie and series production.

Netflix is a subscription-based, Video On Demand (VOD) streaming service for original and acquired movies and series.

Netflix's approach to data-driven decision-making is grounded in three key principles:

1. **Accessibility**—Data and information are readily available to all relevant employees, company-wide.
2. **Visualization**—Data and information are visual and easy to interpret, enabling employees who are not data analysts to understand and utilize the information effectively.

3. **Immediacy**—Access to data and information is immediate, allowing for current information to facilitate real-time decision-making.

The resulting data-driven culture at Netflix marked a significant departure from conventional series production methods and laid the foundation for the enormous success of several hit series and movies.

5.1.6 Data-driven Culture Retention Drill

1. What best describes a company's data culture?

    (a) Financial investments in data analytics tools and technologies.
    (b) Having a dedicated team of data scientists working in isolation from other departments.
    (c) Collective mindset and behaviors within a company that prioritizes the use of data in decision-making processes.
    (d) Data access is restricted to only upper management to ensure confidentiality.

2. Regarding data culture, which is not an attitude toward data?

    (a) Data Denial
    (b) Data Indifferent
    (c) Data Informed
    (d) Data Static
    (e) Data-driven

3. Which six essential steps are required to build a corporate data-driven culture?

    (a) Executive Support and Example
    (b) Quantifying Almost Everything
    (c) Selecting Simple Metrics
    (d) Investment in Database Technology
    (e) Easy Access to Information
    (f) Visualizing Information
    (g) Developing Data Literacy

4. Which three key principles were the basis for Netflix's approach to data-driven decision-making?
    (a) Confidentiality
    (b) Visualization
    (c) Immediacy
    (d) Accessibility
    (e) Intuition

Answers: 1-c, 2-d, 3-abcefg, 4-bcd.

Afterword

The growing reliance on data for making business and product decisions requires product managers to develop and maintain data literacy as a core skill.

The data-driven decision-making (DDDM) process demands that product managers prioritize decisions based on quantitative data instead of intuition or guesswork, understand core concepts related to data, learn to analyze and leverage data, make data-driven decisions, and communicate data-driven decisions to executive management and engineering.

This book is an introductory primer and reference guide for product managers engaging in data-driven decision-making (DDDM).

DDDM Glossary

Acquisition A step in the AARRR framework that is focused on understanding the sources from where users come.

Activation A step in the AARRR framework focused on assessing the quality of a user's initial experience with the product.

Active Users An engagement metric that measures the number of unique and active individuals interacting with the product within a specific time period.

Adoption A metrics category focused on a user's first-time use of a product feature or product.

Adoption Rate An adoption metric that presents the percentage of users who interacted with the product or a specific product feature for the first time within a time frame.

Analytical Dashboards A data dashboard type that investigates data, analyzing and presenting trends, predictions, and insights, thus enabling more informed and predictive decision-making processes.

Apophenia (Patternicity) The human propensity to seek patterns in random data.

Average Revenue Per User (ARPU) A revenue metric in the AARRR framework that represents the average revenue generated from each user or customer within a specific period. ARPU is calculated by dividing the total revenue from all users by the total number of users.

Baseline The initial value of a variable that serves as a basis for comparison.

Big Data Vast datasets of immense size and complexity.

Bounce Rates An acquisition metric in the AARRR framework that represents the percentage of users who hastily leave the product with minimal or no interaction with the product.

Case Studies Analyzing the activities of an individual, group, or company.

Causation A relationship between cause and effect between two variables, demonstrating a direct and consistent effect of one variable on another.

Churn Rate A retention metric in the AARRR framework that represents the percentage of registered users who entirely cease using the product.

Click-Through Rate (CTR) An acquisition metric in the AARRR framework that represents the percentage of users engaging with the product's call-to-action feature.

Cobra Effect An incentive program or policy that results in an opposite target goal.

Communicating Information The strategic presentation of data insights, analysis results, and recommendations in a clear and comprehensible manner.

Comparing Measuring the similarity or dissimilarity between two entities.

Comprehensive Data Software (Do It All) Comprehensive data software solutions that offer an integrated data management approach, encompassing data analysis, visualization, and dashboard features.

Consumption A metric that describes the extent to which users utilize a product, reflecting their usage intensity.

Conversion Rate (Acquisition) An acquisition metric in the AARRR framework that represents the percentage of users who perform any desired action.

Conversion Rate (Activation) An activation metric in the AARRR framework that represents the percentage of users who perform a specific action, such as registering.

Correlation Quantification (measurement) of the relationship or connection between two or more variables. A correlation coefficient is a number between −1 and 1 that indicates the strength and direction of the relationship.

Cost Per Click (CPC) An acquisition metric in the AARRR framework that represents the price paid for each interaction in online advertising campaigns.

Crafting Questions A step within the DDDM process focused on designing clear, relevant inquiries to guide data exploration.

Customer Acquisition Cost (CAC) An acquisition metric in the AARRR framework that represents the expenditure to attract new customers through various channels.

Customer Lifetime Value (CLV) A revenue metric in the AARRR framework that represents the predicted total amount of money a user will bring to the business over their entire engagement with the company.

Data A collection of unorganized facts. Data, whether numerical or textual, lacks any specific meaning or context.

Data Analysis A process that systematically examines, transforms, and arranges datasets to extract patterns and insights from data and convert data to information.

Data Analysis Software (Statistics) Specialized software applications that perform statistical analyses on datasets.

Data Analyst A technical role that requires proficiency in various analytical tools, statistical methods, and data visualization techniques.

Data Collection The systematic process of gathering data from different sources to answer specific research questions, test hypotheses, and evaluate outcomes.

Data Collection (DDDM) A step within the DDDM process focused on gathering raw data from different sources.

Data Culture The collective mindset and behaviors within a company that prioritizes the use of data in decision-making processes. Data culture encompasses the values, practices, and standards that guide how data is collected, managed, and utilized across all company levels.

Data Dashboard Software (Real-Time Metrics) Specialized software applications focused on displaying real-time metrics and Key Performance Indicators (KPIs).

Data Dashboards (Data Visualization Software) Software applications that graphically represent data and information in real-time or as part of an analysis. Data dashboards transform textual and numerical data into visual charts, figures, and tables, making the data and the resulting information more accessible and understandable for users.

Data Denial Attitude People who distrust data and actively avoid its use in their decision-making processes.

Data Imputation A core task within data preparation focused on replacing missing data values with substituted values using mean, regression, or multiple imputations.

Data Indifferent Attitude People who lack interest or concern for data.

Data Informed Attitude People who use data to support their intuition and decision-making.

Data Integration A core task within data preparation focused on combining data from various sources while ensuring that the integrated data is coherent and free from redundancy.

Data Inventory A systematic process focused on cataloging and organizing all data assets by identifying, documenting, and classifying data sources, types, formats, and storage locations.

Data Inventory (DDDM) A step within the DDDM process focused on cataloging available data sources.

Data Literacy The capacity to read, manipulate, analyze, and communicate data.

Data Mining A process facilitated by specific algorithms and dedicated to extracting valuable information from extensive databases.

Data Preparation The meticulous and systematic transformation of raw data into a format suitable for analysis by removing or correcting incorrect, incomplete, or irrelevant data.

Data Preparation (DDDM) A step within the DDDM process focused on cleaning, transforming, and structuring data for analysis.

Data Science The application of data to solve real-world problems about companies and products.

Data Science (Formula) {Data science} = {data mining} + {computer science}

Data Transformation A core task within data preparation focused on converting data into a consistent format, which may include standardizing units of measurement, normalizing text data, and encoding categorical variables.

Data Validation A core task within data preparation focused on checking for and rectifying errors such as invalid entries, outliers, and discrepancies.

Data Visualization A step within the DDDM process focused on visual/graphical representations of numerical statistical figures.

Data Visualization Software (Visuals) Specialized software applications designed to transform raw data into visual formats such as charts, graphs, and maps.

Data-Driven Attitude People who heavily rely on data to shape their decisions.

Data-Driven Decision-Making (DDDM) A systematic process that leverages data and statistical analysis to identify patterns and gain insights that support business and product decisions.

Decisions Judgments or choices that instigate action based on acquired knowledge. Decisions are determinations reached after consideration.

Depth Of Use An engagement metric that measures the quality and richness of user interactions beyond surface-level engagement.

Descriptive Query (Past) A type of DDDM query that only presents facts and describes what has happened.

Detecting Consistencies And Inconsistencies A core task within data analysis focused on ensuring data integrity is crucial for accurate analysis. This task involves identifying and resolving discrepancies or anomalies in the data to maintain consistency and reliability.

Diagnostic Query (Past) A type of DDDM query that presents facts plus an interpretation to provide context, explaining why something happened.

Dwell Time An acquisition metric in the AARRR framework that represents the duration of the user's session with the product.

Engagement A metrics category that describes the level of user interaction and activity with the product, often measured by metrics like daily active users or time spent.

Engagement Metrics Variables that measure ongoing interaction with a product feature or product.

Event An online tracking specification component that describes a single user interaction with a digital product that is being tracked. Examples of events in a website include clicks, hovers, scrolls, touches, app openings, and mouse cursor positions.

Event Properties An online tracking specification component that describes additional data points for the user interaction event. Event properties capture the context around an event. Event properties in a website include page URL, time, date, browser, and device.

False Correlation A statistical phenomenon where two variables appear to be related, but in reality, they are not.

False Negative Error An erroneous condition that occurs when an existing pattern is undetected. This error means a present relationship between data variables goes undiscovered.

False Positive Error An erroneous condition that occurs when a non-existent pattern is incorrectly detected. This error means assuming a relationship between data variables when there is none.

Focus Groups A qualitative data collection method where a trained moderator leads a guided group discussion to gather insights on participants' perceptions, opinions, and attitudes towards a specific topic.

Fundamental Statistics The basic principles and techniques of organizing and analyzing data.

Information Organized data. Information has meaning and context that can be useful and can be understood.

Information Visualization The process of translating information into visual formats such as graphs, diagrams, tables, images, and charts. Information visualization depicts different types of information that support decision-making: trends, comparisons, relationships, and revelation data.

Interviews A qualitative data collection method involving direct interaction between an interviewer and interviewee to gather detailed information on thoughts, experiences, and perceptions.

Knowledge The capacity to comprehend and utilize information, forming the foundation for making decisions. Knowledge encompasses the insights and conclusions derived from the information.

Linear Regression A statistical technique to estimate the relationship between an outcome variable (the data of interest) and a predictor variable (the data used to forecast the outcome variable).

Making Decisions Making informed product and business choices that are guided by data-driven insights.

Market Share A metric that describes the portion of a market that a company's product captures relative to competing products.

Mean (Average) The central value of a numerical dataset, calculated by summing all the numbers in the dataset and then dividing the sum by the total count of values.

Median The middle value in a sorted dataset. The middle value splits a sorted dataset into two equal parts.

Metadata Data that describes other data, but it is not the other data itself.

Metrics Selection A step within the DDDM process focused on choosing specific performance indicators for evaluation.

Mode The most frequent number in a given dataset.

Negative Correlation The variables, related to each other, change in opposite directions.

Net Promoter Score (NPS) A referral metric in the AARRR framework that represents the probability that users will recommend the product to others.

New Product Features Used by Existing Users A specific Time to First Action (TFA) adoption metric that presents the mean time for the first use of new product features by existing users.

New Users Using Existing Product Features A specific Time to First Action (TFA) adoption metric that presents the mean time for the first use of existing product features by new users.

North Star Metric A product variable the entire company focuses on to achieve long-term growth.

Observations A data collection source based on systematically watching and recording users' behaviors, events, or conditions in their natural setting in real-time.

Online Data Data collected from social media platforms, forums, or online communities to understand views and opinions.

Online Tracking The act of constantly collecting and storing user activity data with a digital product. Online tracking involves collecting data on users' behaviors and interactions within digital environments using technologies like web analytics, cookies, and tracking pixels.

Online Tracking Specification A structured description of the user interactions and associated data points to be tracked. The online tracking specification consists of three components: event, event properties, and user properties.

Open Data A social concept of making data freely available to the public to use and share without restrictions from copyright, patents, or regulatory constraints.

Operational Dashboards A data dashboard type that displays key metrics in real-time, providing immediate insights into current performance and allowing for prompt decision-making and action.

Pattern Recognition A core task within data analysis focused on identifying recurring patterns or regularities within the dataset is essential for making predictions and identifying significant trends. Pattern recognition techniques enable the detection of meaningful structures within the data.

Positive Correlation Both variables, related to each other, change in the same direction.

Predictive Query (Future) A type of DDDM query that foresees the future according to the facts and describes what will happen.

Prescriptive Query (Change The Future) A type of DDDM query that attempts to alter the future, questioning how to make something happen and what will happen if a certain action is taken.

Purchase Rate of Referred Customers A referral metric in the AARRR framework that represents the percentage of referred users that become paying customers.

Qualitative Data Interpretive and observational facts which provide descriptive qualities and characteristics that are not quantifiable, defined, or confined by numbers.

Qualitative Data Analysis An interpretive process that provides insights into experiences, perceptions, and behaviors.

Quantitative Data Facts characterized by their numerical and factual nature and expressed using numbers.

Quantitative Data Analysis A mathematical process that emphasizes numbers, statistics, and other numerical elements, such as the median and standard deviation.

Query Format and Components *"As a product manager, I want to know [**question**] because [**objective**], so I need [**data**]."*

Query Formulation A step within the DDDM process focused on creating precise requests for data retrieval and analysis.

Questionnaires A data collection tool consisting of questions designed to gather information from respondents.

Referral A step in the AARRR framework focused on turning existing users into advocates who recommend the product to others through referral programs, incentives, and social sharing product features.

Referral Rate A referral metric in the AARRR framework that represents the percentage of existing users who refer others to the product.

Relationship The existence of a correlation or causation between variables.

Retention A step in the AARRR framework focused on tracking how many users are retained over time and understanding why some leave. The goal is to build loyalty and reduce churn.

Retention Rate A retention metric in the AARRR framework that represents the percentage of users who continue using the product over time.

Revelation Data Disclosing previously unknown facts.

Revenue A metric that describes the total money a business earns from selling its products.

Revenue (AARRR) A step in the AARRR framework focused on devising strategies to increase overall revenue and monetize the user base through pricing optimization, upsells, or expanding the user base.

Revenue Growth Rate (RGR) A revenue metric in the AARRR framework that represents the increase in a company's total revenue over a specified period, typically expressed as a percentage.

Statistical Calculations A core task within data analysis focused on applying statistical methods to analyze data distributions, relationships, and variations.

Strategic Dashboards A data dashboard type that summarizes key metrics over a specified period, offering a comprehensive overview that supports long-term planning and strategic decision-making.

The AARRR Framework A 5-step framework to help select the appropriate metrics for growth for startups and digital products. The AARRR acronym stands for acquisition, activation, retention, referral, and revenue.

Time Spent An engagement metric that measures the number of sessions and the duration the users actively spend with the product.

Time to First Action (TFA) An adoption metric that presents the mean time for new users accessing an existing product feature or existing users accessing a new product feature for the first time.

Time to Value An activation metric in the AARRR framework that represents the mean time for users to recognize the product's benefits.

Traffic Source An acquisition metric in the AARRR framework that represents tracking and listing the origin and ratio from where users came to the product.

Trend The general direction where something is changing or developing.

Usage Frequency An engagement metric that measures how often users return to engage the product, reflecting habitual behavior.

Usage Recency An engagement metric that measures the elapsed time since the user's last interaction with the product.

User Properties An online tracking specification component that captures the context around the user performing the event. Examples of user properties related to a website event include the user's gender, age, location, and user category (registered or unregistered, paid or free).

User Satisfaction A metric that describes the level of positivity or negativity that users perceive in their interactions with a product, often measured through Net Promoter Score (NPS) or other feedback mechanisms.

Viral Cycle Time A referral metric in the AARRR framework that represents the mean time it takes a user to refer others to the product.

Visitors to Registration Ratio An activation metric in the AARRR framework that represents the percentage of users registering for the product. Also, specifically in response to a marketing effort.

Visual Data Analyzing images or videos to understand people's experiences or perspectives.

Zero Correlation No linear relationship exists between the variables.

Printed in the United States
by Baker & Taylor Publisher Services